I0625747

FEARLESSLY USE AI IN THE CLASSROOM

EASY-TO-USE PROMPTS, DEMYSTIFY AI WITH A STRESS-FREE GUIDE TO ESSENTIAL STRATEGIES AND STEP-BY-STEP TECHNIQUES FOR TRANSFORMATIONAL CLASSROOM INSTRUCTION

MICHAEL P. WEST

Published by:

Ace East Publishing LLC

1401 21st Street Suite R

Sacramento, CA 95811

ISBN: *978-1-963939-18-7*

ASIN: [Insert ASIN]

Cover Design & Interior Design, and Formatting by
Ace East Publishing
Printed in the United States of America

ACE EAST
PUBLISHING

Disclaimer:

This is a work of nonfiction. Names, characters, places, and incidents either are the product of the author's imagination or are used fictitiously. Any resemblance to actual persons, living or dead, events, or locales is entirely coincidental. The publisher and author make no representations, warranties, or guarantees as to the accuracy, completeness, or suitability of the content provided in this book.

For more information, visit aceeastpublishing.com.

First Edition

November 20, 2024

*To **John Vafis**,*

A pioneer, an innovator, and a mentor.

Your unwavering vision brought the world of computers into the classroom before many even knew what was possible. From the days of real-to-reel programming to a time when floppy disks were famous, you paved the way for technology to take root in our lives.

As a former instructor at Colusa High School, you not only introduced an entire generation to the future but also inspired us to embrace curiosity, creativity, and endless possibilities. Your teaching, wisdom, and leadership ignited a spark that continues to burn bright in the lives of those you guided.

Thank you for being more than an educator. Thank you for being a trailblazer, a mentor, and a true role model. Your legacy will live on through every student you inspired, including me.

With the deepest respect and gratitude,
Michael West

CONTENTS

HERE'S MY PROMISE TO YOU

To jumpstart your journey and make AI work for YOU, I'm thrilled to offer you an exclusive gift:

A FREE Listing of 1,000 Prompts You Can Use in Your Classroom Starting Tomorrow!

That's right, you now have access to an entire treasure trove of ready-made prompts to inspire creativity, build critical thinking, and streamline your teaching workflow. Whether it's brainstorming, research, writing assistance, or even lesson planning, these prompts will empower you to **get the most out of your GPT AI tools**.

CLICK HERE TO DOWNLOAD YOUR FREE PROMPT LIST

ENJOY, AND GET TO PROMPTING!

The possibilities are endless, and the power of AI is now at your fingertips. Let this book and your FREE prompt list be the springboard to creating a transformational classroom experience.

Thank you for your trust in this guide, and here's to fearless teaching with AI, your students will thank you for it!

Now, go ahead and start prompting your way to brilliance. The future of education is in your hands!

You've got this!

INTRODUCTION

THE FUTURE IS NOW: EMPOWERING EDUCATORS TO TRANSFORM LEARNING WITH AI

Imagine walking into a classroom where students collaborate seamlessly with artificial intelligence to solve complex real-world problems, paint masterpieces, write poetry, or explore the far reaches of historical inquiry. Picture students actively critiquing AI-generated biases, developing ethical awareness, and shaping their own original ideas, all while using cutting-edge tools as co-creators, not competitors.

The AI revolution is no longer coming; **it's already here**. For educators, this presents an extraordinary opportunity. Artificial intelligence has the power to **amplify creativity, deepen critical thinking**, and **personalize learning** in ways that were once the stuff of science fiction. But it also brings challenges, ethical dilemmas, concerns about originality, and the urgent need to redefine how we teach collaboration, problem-solving, and creativity. How can we ensure that AI enhances education rather than disrupts it? How do we maintain the essential human elements of learning, curiosity, empathy, and originality while leveraging the incredible potential of AI?

This book answers those questions and more, providing educators with a **practical, step-by-step guide** to integrating AI into their classrooms thoughtfully, ethically, and effectively. From **STEM** to the **arts and humanities**, this book will show you how to harness the power of AI to inspire, engage, and empower students in ways that feel both futuristic and deeply human.

Here's what you'll discover:

- **Transformative AI Strategies for the Classroom**: Learn how to design collaborative, AI-supported problem-solving activities that engage students, promote teamwork, and tackle real-world challenges.
- **Revolutionizing Creativity and Critical Thinking**: See how AI can inspire students to push the boundaries of art, writing, and design while preserving their unique voices and fostering originality.
- **Prompt Engineering Mastery**: Unlock the art and science of creating effective AI prompts that guide students toward clarity, creativity, and critical reflection.
- **Practical Tools and Real-World Scenarios**: Explore tangible examples, worksheets, and adaptable frameworks that make AI integration seamless for every grade and subject.
- **Ethics and Equity**: Discover strategies to address bias, ensure inclusivity, and promote ethical AI literacy, empowering students to engage thoughtfully with the technology shaping their future.

This is not just a book about **technology**, it's a book about **teaching for the future**, preparing students to thrive in an AI-driven world where critical thinking, creativity, and ethical decision-making matter more than ever. It is a guide for educators who refuse to be left behind and who see AI not as a threat but as an opportunity to **redefine what's possible** in education.

Whether you're an experienced tech-savvy educator or someone just beginning to explore AI's potential, this book provides the tools, insights, and confidence you need to unlock the power of AI in your classroom.

The future of education is collaborative, creative, and transformative. Let this book be your guide as you lead your students into that future, one where human ingenuity and artificial intelligence work hand in hand to create something greater than either could achieve alone.

Are you ready to unleash the possibilities? The classroom of tomorrow starts today.

1

UNMASKING AI

BREAKING DOWN THE BASICS FOR EVERY EDUCATOR

OVERHEARD IN THE STAFF ROOM

"I just don't see how AI can ever replace what we do," Ms. Johnson overheard her colleague say in the staff room. "But imagine how it could help us reach every student," another replied. Ms. Johnson listened intently, a sense of uncertainty washing over her.

As a passionate educator with years of experience, Ms. Johnson has dedicated her career to connecting with students and honing her craft. But now, with AI poised to transform education through adaptive learning, personalized tutoring, and automated assessment, she wondered about her role in this new landscape.

Some of her colleagues were excited, envisioning AI as a tool to personalize learning and engage students like never before. Others were skeptical and worried about AI's implications for their jobs and its potential to worsen inequalities. Ms. Johnson couldn't ignore these developments, but the jargon was confusing, and the implications overwhelming. She wished someone would cut through the hype and clearly explain how to harness AI's potential.

Standing at this crossroads, Ms. Johnson felt torn. She could stick with her tried-and-true methods, the lessons she'd spent years perfecting. Or she could step into an uncertain but exciting future where AI might become a powerful ally in helping every student succeed. If you've ever found yourself in Ms. Johnson's shoes, caught between the comfort of the familiar and the pull of innovation, this book is for you.

In the following chapters, we'll be your expert guides to understanding AI's educational impact. We'll demystify the jargon, separate fact from fiction, and equip you with a roadmap for confidently implementing AI in your classroom. Whether you're a technophile or a skeptic, a veteran teacher, or a new educator, you'll gain the knowledge and strategies to thrive in the age of AI.

Get ready to explore real-world applications, gain practical insights, and discover how to harness these powerful tools to support your students. As we dive into the world of AI, it's essential to understand that the term "Artificial Intelligence" has become ubiquitous in our daily lives, often used casually in various contexts.

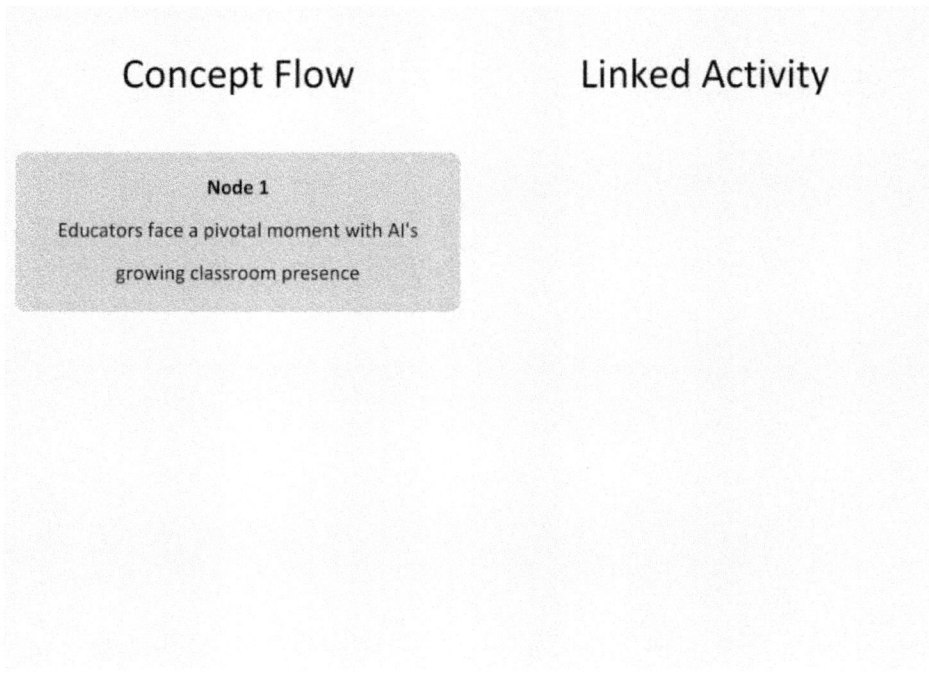

WHAT IS ARTIFICIAL INTELLIGENCE? A SIMPLE EXPLANATION FOR NON-TECHIES

Artificial Intelligence (AI)is more than just a buzzword. Its prevalence in our lives is so significant that you might think you already know how it works. However, when it comes to AI in education, its application and understanding can be quite different. Let's unravel AI's potential to improve student outcomes.

AI is already here, and it represents a significant branch of computer science dedicated to crafting machines that emulate tasks requiring human intelligence. When you're on the sofa watching Netflix, which uses AI to analyze your viewing history, identify your preferences, and provide personalized recommendations, that's AI in action. Netflix's AI creates micro-genres from your viewing patterns, offering specific content suggestions. This individualized service shows AI's role in enhancing your digital engagements, turning a streaming site into an intelligent content curator. Without AI, your streaming experience would be far less tailored, potentially leading to endless scrolling and less satisfaction. Similarly, AI can analyze your student data, identify your learning patterns, and provide solutions in education.

Imagine if every student in your classroom could receive personalized learning experiences tailored to their specific needs and learning styles. By utilizing AI algorithms, platforms like Khan Academy can identify each learner's gaps in knowledge and deliver targeted educational resources to help them grasp challenging concepts and fill those gaps.

For example, Brian did poorly in math, but his teacher could not discern why. She used AI to determine that he didn't truly understand how to apply the concepts to real-life scenarios. The AI-powered platform provided him with targeted practice problems and real-world applications, helping him bridge the gap between theory and practice. His understanding grew, boosting his confidence and performance in math significantly.

Like Netflix, AI turns Khan Academy into an intelligent learning management system instead of just a generic learning site. This is just one example of how AI is revolutionizing the educational landscape by making learning more engaging but also about making it more effective.

DECODING AI JARGON: ALGORITHM VS. MACHINE LEARNING VS. DEEP LEARNING

As you venture into AI, you'll encounter terms like Machine Learning and Deep Learning. These terms, while related, are different, and understanding the nuances is critical for you to appreciate the advancements in this field.

At a glance, AI is a broad concept that includes all types of machine intelligence that can carry out tasks in a manner you would consider "smart." It involves researchers and scientists creating algorithms that enable machines to carry out tasks that typically require human intelligence, like learning, reasoning, problem-solving, understanding language, and perception.

You might be wondering, what exactly is an algorithm for AI in the classroom?

An algorithm is a set of rules or instructions designed to perform a task or solve a problem. At its core, an algorithm is a step-by-step instruction.

In a field as broad as AI, it's often easier to describe something by what it is not rather than what it is. For example, when pressed, you might argue that AI is an "independent-thinking robot," suggesting that any AI model can do any task. However, this is a misconception. In reality, AI systems are designed for specific tasks and operate under heavy constraints.

To clarify, *Machine Learning is a specific application within AI, while Deep Learning is a more specialized version of Machine Learning designed for handling complex tasks.* By understanding these distinctions, we can better grasp the nuances of AI and its various subfields.

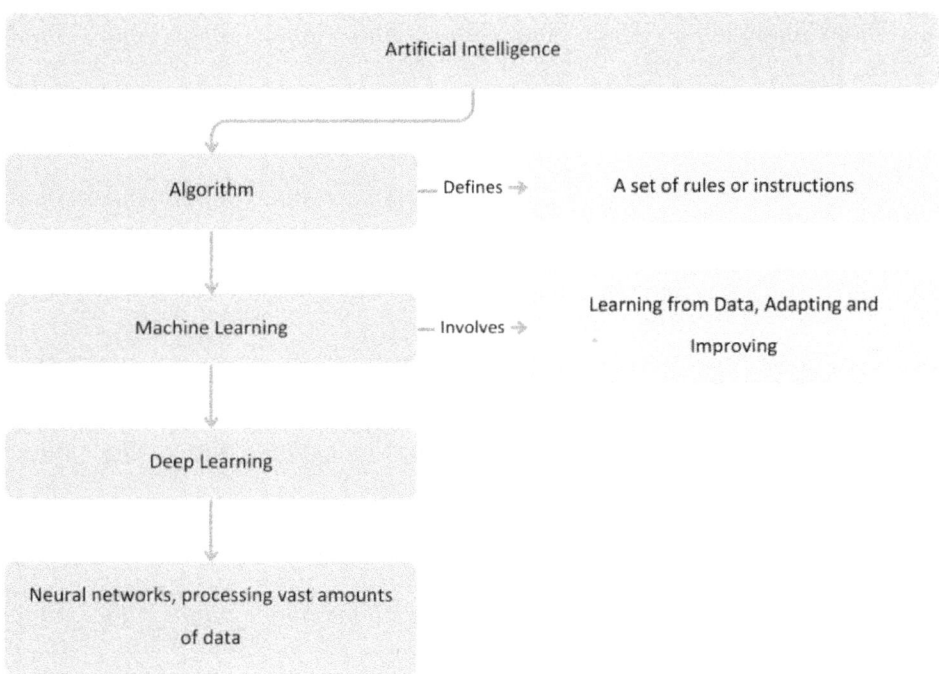

Machine Learning, you'll discover, is all about developing algorithms for machines to learn and make decisions based on data. Unlike traditional programming, where you code the rules, Machine Learning algorithms adapt and improve based on the data.

What does this look like?

Let's consider your own experience. When you shop online, Machine Learning enhances product recommendations based on browsing and purchasing history. These algorithms continuously learn from your interactions and update real-time recommendations to offer a more personalized shopping experience.

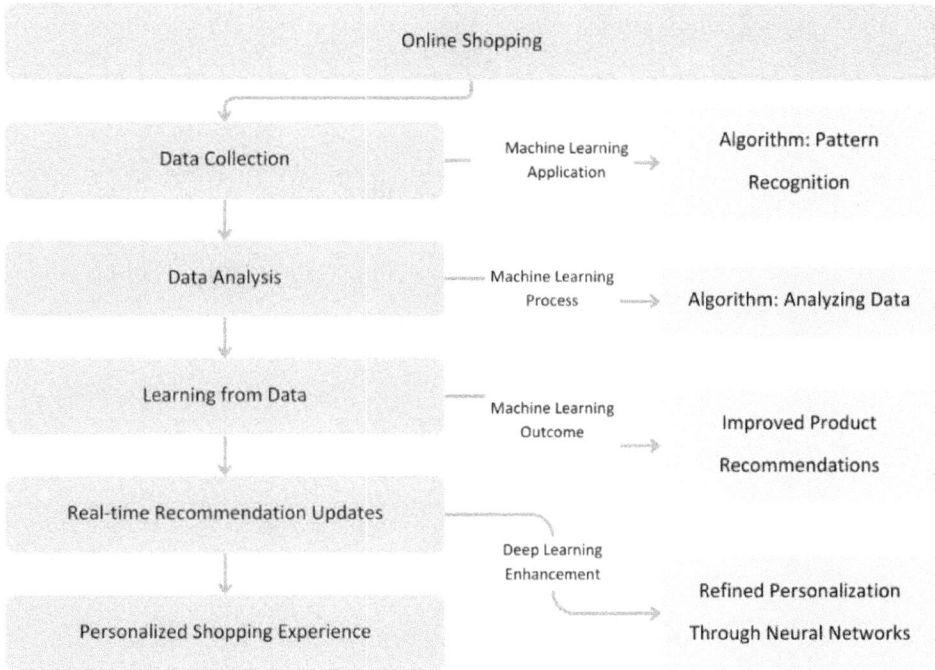

Machine Learning starts with simple algorithms that improve over time as they process more data. For example, the longer you watch Netflix, the better the recommendations become. This isn't because the algorithm is getting more complex; it's because it's processing more data from your viewing habits, discovering patterns, and refining its predictions.

Deep Learning goes beyond Machine Learning by using neural networks that mimic the human brain. These neural networks, which are composed of layers of interconnected nodes, help handle and interpret vast amounts of data.

When we refer to the data needed, we talk about enormous data sets, terabytes, and more information. In technical terms, Deep Learning uses networks with many layers (hence 'deep') to process data.

Do you ever think about how you make decisions? It's easy to say Deep Learning uses networks, but many of us are unaware of how we make decisions. So, it's hard to understand Deep Learning without understanding it.

Let's take a simple decision: how do you decide what to make for lunch?

As you approach the first layer, you're dealing with your initial, basic considerations. You think about immediate factors like what ingredients you have on hand, how much time you have to prepare or eat, and what you're in the mood for. This is similar to the initial layer in a Deep Learning model, which processes the most basic and obvious features.

As you delve deeper, consider more complex factors, like nutritional balance based on your light breakfast or need for a substantial lunch. In Deep Learning, this is like intermediate layers that extract complex patterns and relationships and understand context or abstract features.

Then there are the subtle, often subconscious influences. Perhaps the gloomy weather makes you crave comfort food, or a conversation about Italian cuisine triggers your desire for pasta. Similarly, looking at a Deep Learning model is like peering into a system that analyzes the data's abstract and nuanced aspects. These layers understand sentiments, complex relationships, and intricate patterns not immediately obvious.

As you consider what to make for lunch, imagine three interconnected nodes that lead to your final decision (also called the "output"):

1. Initial considerations: assessing available ingredients and time constraints.
2. Intermediate considerations: factoring in nutritional needs based on previous meals.
3. Deep considerations: taking into account mood and external influences.
4. Final decision: choosing a meal based on the cumulative analysis of all factors.

Deep Learning models process information, starting with basic data and gradually extracting more complex insights. This progressive refinement leads to an informed output or decision, like how you mimic human decision-making processes in artificial intelligence.

Ultimately, you combine all these layers of thought to decide what to have for lunch. This decision results from obvious and subtle considerations, processed at different levels of your thinking.

In Deep Learning, the final output, classification, prediction, or content generation results from processing through multiple layers. Each layer adds complexity and depth to the understanding of the input data, like your lunch decision, which results from multiple layers of thought and consideration.

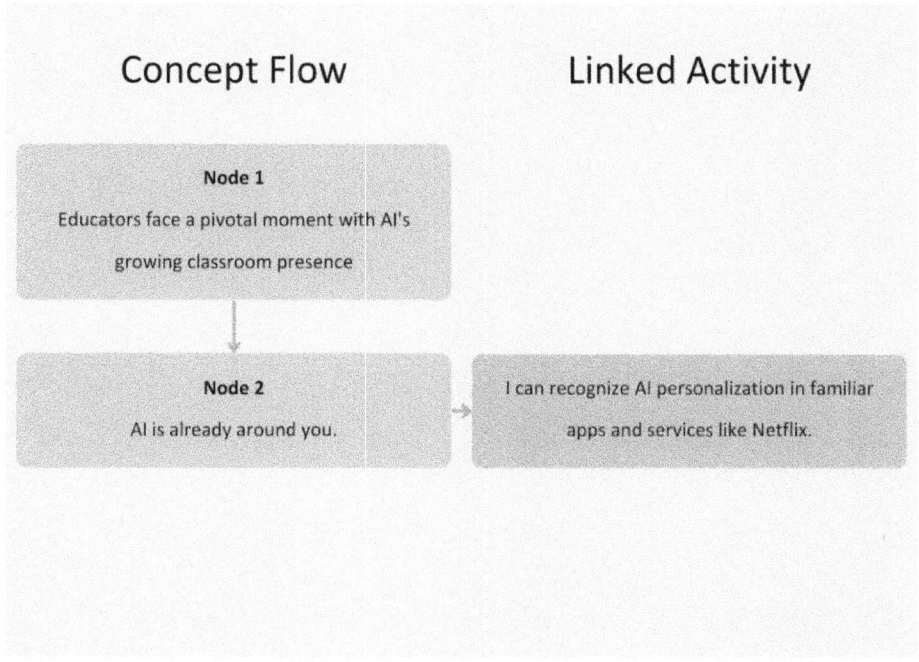

SIDEBAR: Natural Language Processing (NLP)

We can't talk about AI without knowing what Natural Language Processing (NLP) is. The term is often used as a distinct term from machine learning, even though NLPs rely heavily on machine learning algorithms to understand and generate language over time.

NLP is a field of AI that concentrates on teaching computers to comprehend, decipher, and respond to human language in a meaningful and beneficial way. Its goal is to bridge the gap between human communication and computer understanding. When you chat with ChatGPT, for example, you interact with an advanced NLP model. It interprets your typed words, understands the context, and generates a suitable response. This is why it can carry on a conversation that feels almost human.

The core components of NLP are syntax, the structure of language, and semantics, which pertain to the meaning of words and phrases. Syntax involves grammar, sentence structure, and word arrangement. At the same time, semantics is concerned with interpreting the meaning of words and sentences.

Natural Language Processing is about way more than just breaking down the grammar of sentences. It has to truly understand the meaning behind the words. Take this example: "The student reads a book." NLP doesn't just see "student" as a noun, "reads" as a verb, and "book" as another noun. It knows that "student" refers to a learner, "reads" means consuming written material, and "book" is a published work. But it gets even deeper than that. NLP uses context clues to determine whether we're discussing reading for fun or a class assignment. The same sentence could have different implications. It also taps into vast knowledge databases to disambiguate the intended meaning and make connections to real-world concepts. On top of that, NLP can reason and infer things beyond the literal words. From this example, it might be deduced that if students read, they likely have access to educational resources and are actively learning. So, while understanding the grammar is step one, NLP employs semantics, context analysis, reasoning, and background knowledge to achieve true comprehension, just like humans do.

SIDEBAR END

New Keywords: Adaptive Learning, Data Analytics, Virtual Reality (VR), Augmented Reality (AR), Mixed Reality (MR), Learning Management Systems (LMS)

 *I'm not convinced it's anything more than a passing trend. I'll wait and see if it stands the test of time...I want to see the data on how other teachers integrate AI into their classrooms before considering its relevance to my teaching practice. And, more importantly, I want to see the assessment data of our students.*

— MR. D, SOCIAL STUDIES TEACHER

Mr. D's skepticism is understandable. As with any emerging technology, educators want clear evidence that AI can sustainably enhance student learning outcomes before fully embracing it. While initial studies and pilot programs show promise, more longitudinal data and proven examples are needed to demonstrate AI's lasting impact. As adoption grows, gathering and sharing data on best practices, challenges, and measurable outcomes will be required to inform evidence-based implementation.

However, current studies and pilot programs provide some initial insights, which we will explore further in this guide.

You can often do more than one thing with a tool in technology. In comparison, an everyday object like a pen allows you to write but lacks an editing function.

When you're exploring technology tools, you'll find that it's often a case of "It does this, AND it does that." So, while we can list the standard ones, they often span multiple functions. It might be more helpful to define the normal tasks of AI tools in education rather than listing them. Also, everything is changing rapidly. What may be a market leader now might not hold that position tomorrow.

PERSONALIZED LEARNING TOOLS

Personalized Learning Tools combine AI with advanced data analytics. The primary goal is to adapt teaching methods to what you may have known for centuries: You have unique learning preferences, abilities, and educational needs and will learn best if they are met.

Personalized approaches recognize the limitations of a 'one-size-fits-all' approach. Given the class size, a teacher can't customize the learning process for each student without assistance.

DreamBox Learning uses a combination of adaptive math and reading instruction to help students improve their skills. The platform adjusts the material's difficulty based on the student's performance, providing more challenging problems when the student is ready and fewer when the student needs additional practice.

During the 2022-23 academic year, the Salem School District implemented DreamBox Math for students in grades K-8. The program required all students to complete a minimum of five lessons per week. The school also organized a math challenge in May 2022 to further engage students and foster a sense of friendly competition.

Craig Velleux, the data specialist who implemented the program, saw remarkable results. Students using DreamBox regularly, especially third-grade students, showed improved Wisconsin Forward scores.

Source: https://www.dreambox.com/resources/case-study/salem-school-district-students-who-used-dreambox-math-regularly-improved-map-scores

Another example of a personalized learning tool is using AI-powered tutoring systems like ChatGPT. These tools can offer personalized assistance and resources based on each student's behavior, similar to a tutor bot like Duolingo. These systems can provide tailored lessons, practice, and feedback by tracking students' progress, strengths, and weaknesses, making the tutoring more targeted and effective. Automated tutors can provide unlimited patience and support that human tutors may not always offer.

Key Features:

- Personalized Learning Tools use sophisticated algorithms to create individualized learning paths for each student, considering their unique learning pace and style rather than using a standardized approach for all learners.
- AI algorithms gather data on student performance, including their answers, time spent on topics, and areas of difficulty. This data helps personalize learning materials and pinpoint areas where students need more support.
- The personalized approach optimizes learning by providing content at the right difficulty level and relevance for each individual. This minimizes boredom from overly easy material and frustration from too complex content, keeping learners motivated.

SIDEBAR: Leveraging AI to Generate Personalized Writing Prompts: An Interactive Walkthrough

While Dreambox relies on subscription-based models that need to be implemented on a school or district-wide level, below is an example you can use today to personalize the learning you give to your students.

Create personalized writing prompts for your students' interests with AI language models like ChatGPT and a simple spreadsheet. This demo will guide you through the process and show you how to generate personalized story premises that inspire creativity.

Step 1: Prepare a spreadsheet with student interests

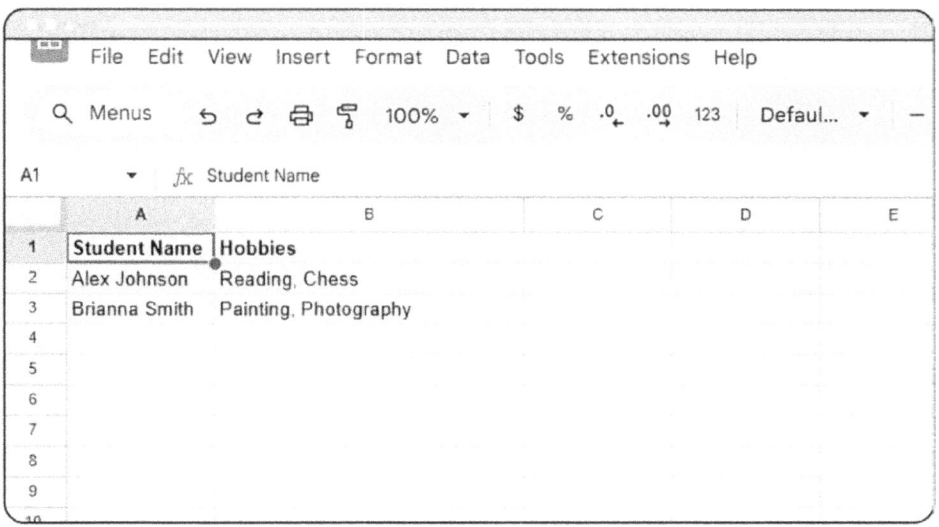

Step 2: Provide Initial Prompt and Data Source

Enter a single prompt into ChatGPT, along with the spreadsheet containing the student interest data from Step 1. The prompt will instruct ChatGPT to reference the spreadsheet and generate a unique story premise for each student, incorporating their hobbies.

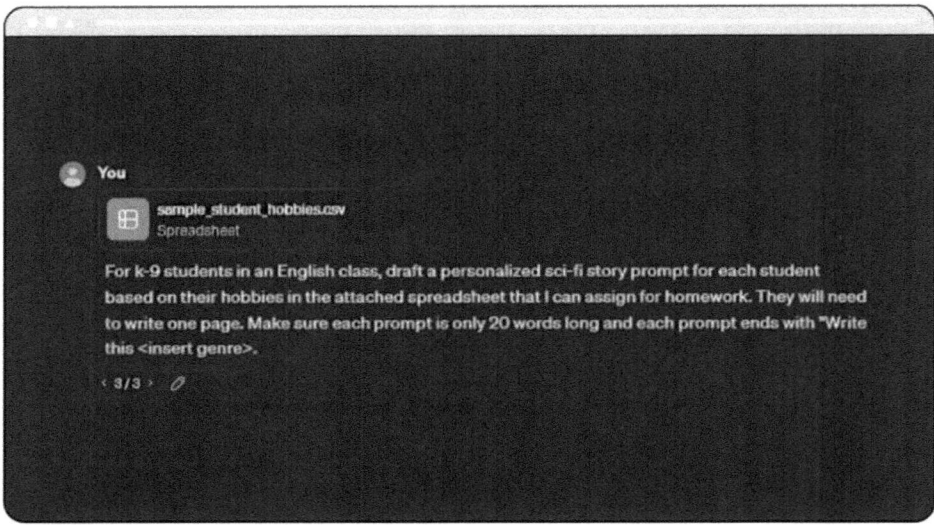

Step 3: ChatGPT Generates Personalized Story Concepts

ChatGPT will process the prompt, access the attached spreadsheet, and generate imaginative sci-fi story ideas for every student on the list, incorporating each student's hobbies and interests.

For example:

Student: Alex Johnson (Reading, Chess): "In a future where books control reality, a chess prodigy must outsmart a tyrannical librarian. Write this sci-fi."

Student: Carlos Rivera (Coding, Robotics): "A teenager codes the first AI robot capable of dreaming, but its dreams predict the future. Write this sci-fi."

Using AI's language generation, you can create diverse writing prompts that resonate with each student's interests with minimal effort.

Step 4: Explore more applications

This personalized prompt generation can be adapted for various subjects and learning objectives. Consider using ChatGPT to create:

- Customized math word problems based on student interests
- Personalized science experiment ideas tied to real-world applications
- Individualized historical fiction scenarios for social studies

Bonus Tip 1: To streamline the process even further, you can create a simple Google survey to gather information about your students' interests, hobbies, and preferences. Once the students have completed the survey, you can download the responses as a spreadsheet and use them as input for generating personalized content with AI.

Bonus Tip 2: When generating personalized content for many students, be mindful of the output limits of ChatGPT. To ensure that ChatGPT generates the best possible content for each student, processing the prompts in smaller batches of around 10 at a time is recommended. This way, the AI can dedicate more 'time' and 'attention' to each prompt, resulting in higher quality and more detailed outputs. While generating all 30 outputs simultaneously is technically possible, the algorithm may need to shorten the prompts to fit within its output window.

END SIDEBAR

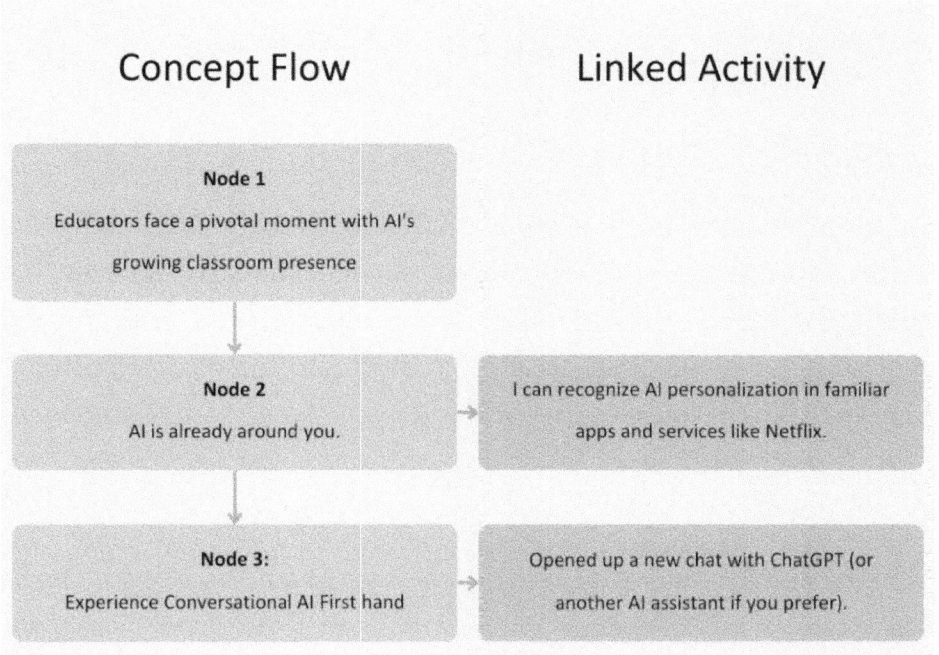

ASSESSMENT AND GRADING TOOLS

Automatic assessment and grading are hardly new in educational technology. Who hasn't added a multiple-choice test to their assessments when the time was short, and they needed a quick grade? But with AI technology, it has taken a much more powerful form. Today, with AI-powered grading tools, teachers can assess students' understanding and comprehension beyond basic multiple-choice formats; they have evolved to evaluate complex responses, including essays and short-answer questions.

One remarkable example is Gradescope, an automated grading software utilizing NLP to grade various data types and subjective answers.

With Gradescope, educators can grade various assignments, including paper-based exams, quizzes, bubble sheets, and homework. It can even grade handwritten student responses.

Source: https://help.gradescope.com/article/h7ztxl9164

Key Features:

- Automated grading saves teachers time and ensures a fair assessment for all students.
- Instant feedback helps students learn from their mistakes and improve their understanding.
- Detailed analytics provide valuable insights into areas of difficulty, allowing teachers to tailor their teaching strategies for better student outcomes.

DATA-DRIVEN DECISION-MAKING TOOLS

Data-driven decision-making tools, or data Analytics, use artificial intelligence to analyze and interpret large volumes of educational data. These tools assist educators and administrators in making informed decisions about curriculum design, teaching strategies, and student support. They provide insights into student performance, learning patterns, and areas of struggle, enabling you to carry out targeted interventions. Moreover, these tools can predict future performance and learning outcomes, helping you shape educational strategies and policies for improved student success.

Learning Management Systems (LMS) such as Canvas or Blackboard aren't new. Still, their implementation of AI has revolutionized how they function. These platforms now incorporate AI to analyze student data, track performance, and provide personalized learning paths. Another example is BrightBytes, a data analytics platform that helps schools and districts understand and improve student learning. It uses predictive analytics to identify students at risk of dropping out, enabling you to intervene early.

Data Analytics Tools like Tableau and Power BI are robust software solutions in various industries, including education. These tools help you make sense of complex data sets and pinpoint trends and patterns that might not be immediately obvious. By transforming raw data into meaningful visualizations, you can gain actionable insights and make data-driven decisions to enhance student outcomes.

Key Features:

- Data analysis in education helps us understand why some students may be struggling more than others by looking at patterns in their performance, attendance, and background information. This information can then be used to make informed decisions and create targeted interventions to support those students.
- By using predictive models and analytics, teachers can identify common learning needs among students and form groups that work well together. This allows students to collaborate and support each other in their learning, making it more effective and enjoyable.
- Data-driven tools provide visual representations of student performance and progress. Sharing these visuals with students, parents, and other stakeholders encourages open discussions about improving and personalizing instruction. Additionally, these tools help identify students who may need extra support early on so that interventions can be implemented to help them succeed.

SIDEBAR: The Difference Between Personalized Learning Tools and Data-Driven Decision-Making Tools

Purpose: Personalized Learning Tools adapt educational content to individual students' needs. Data-driven decision-making Tools provide insights to guide decisions around curriculum, instruction, resources, and policies.

Users: Personalized Learning Tools interact directly with students, while Data-Driven Tools are used by educators, administrators, and policymakers.

Scope: Personalized Learning Tools operate at the individual student level (micro), while Data-Driven Decision-Making Tools analyze system-wide trends and patterns (macro).

Data Use: Personalized Tools utilize individual student performance and learning data. Decision-making tools examine broader institutional datasets, including demographics, resource allocation, and administrative data.

SIDEBAR END

LANGUAGE AND WRITING ASSISTANCE TOOLS

Language and Writing Assistance Tools are AI-driven platforms designed to help students improve their language and writing skills. These tools can offer real-time grammar, punctuation, and style feedback, helping students refine their writing. They can also provide personalized learning resources to enhance vocabulary and language comprehension. These tools are particularly beneficial for non-native English speakers and students who struggle with written communication.

Grammarly and Hemingway Editor are tools that refine user drafts by providing feedback on grammar, punctuation, and style. We are all used to spell-checking, but with AI, these tools take it a step further, giving context-specific corrections and stylistic recommendations, turning it into a learning opportunity instead of just quality control. These language and writing assistance tools help students improve their writing skills by offering real-time suggestions and corrections.

Key Features:

- Powered by AI and Natural Language Processing, they analyze user-written work and offer feedback, error correction, and improvement suggestions.
- These tools improve communication skills for professional and academic success, with a particular emphasis on supporting non-native speakers and students from diverse linguistic backgrounds.
- They allow for collaborative writing and foster connections among language learners, including adaptations for learning disabilities.

IMMERSIVE LEARNING TOOLS

Immersive Learning Tools refer to education technologies designed to create an engaging learning journey using Virtual Reality (VR), Augmented Reality (AR), and Mixed Reality (MR) (see sidebar). The core purpose is to simulate real-world environments or illustrate complex concepts, making learning experiential.

AI is integrated into immersive learning tools to enhance functionality, personalize the learning experience within these environments, adapt the content based on the learner's progress, or even simulate intelligent interactions within VR, AR, or MR.

Immersive Learning Tools equip students with innovative methods to grasp complex concepts. For example, tools like Google Expeditions and Body VR provide educational virtual reality (VR) applications that allow students to go on virtual field trips or explore human anatomy in 3D. These VR experiences enhance learning by providing interactive and engaging opportunities for students to explore and understand subjects in a more hands-on way.

Similarly, augmented reality (AR) learning apps like Elements 4D make subjects like chemistry interactive. Students can use AR technology to blend elements and visualize chemical reactions, making the learning experience more dynamic and enjoyable. These tools utilize AI to create interactive learning experiences, enabling students to grasp concepts more experientially and memorably.

Key Features:

- VR tools create immersive experiences by placing users in a digital world and allowing them to interact with headsets, hand controllers, or treadmills.
- AR overlays digital data onto the real world, enhancing the immediate environment. It can be used with smartphones, tablets, or AR glasses.
- Immersive learning tools engage students with interactive experiences, allowing them to explore complex topics and visualize intricate processes.
- Simulations in virtual environments provide unique learning opportunities, especially for students with physical disabilities. They enable them to overcome physical limitations and engage in otherwise inaccessible experiences.
- Collaborative virtual environments facilitate group activities, discussions, and projects, fostering collaboration and critical thinking skills regardless of physical distance.

SIDEBAR: What is the difference between VR, AR, and MR?

VR (Virtual Reality): This digital experience completely immerses users in a simulated environment. In VR, users wear a headset that tracks their head movements in three dimensions, allowing them to look around the virtual environment as if they were there. Some VR systems also include hand controllers to interact with the environment.

AR (Augmented Reality): Unlike VR, AR overlays digital information on the real-world view. This means users can see and interact with virtual elements superimposed on their surroundings. This technology is commonly used in mobile games, navigation apps, and educational tools.

MR (Mixed Reality): MR combines elements of VR and AR. It places virtual objects in the real world and allows users to interact with them as if they were real.

POTENTIAL CHALLENGES AND HOW TO OVERCOME THEM

Below, we'll provide an overview of the key challenges and concerns surrounding AI in education. Subsequent chapters will explore these issues in greater depth, along with specific mitigation strategies. We've included a quick reference guide to the relevant chapters.

Major technological developments have always met resistance because people fear they will disrupt existing industries and societal norms. When television first arrived, some saw it as a revolutionary invention that would change how we consume information and entertainment. At the same time, many feared it would replace other entertainment forms like radio and cinema and take over the education sector, with TV screens replacing classrooms. However, these fears proved unfounded. Instead of replacing other mediums, television complemented them. Similarly, it did not replace traditional classrooms but became a supplementary education tool.

Today, we face similar fears with AI. It's crucial to understand that AI is not here to replace educators but to enhance their capabilities. Here are the key challenges and concerns with AI in education:

1. Ethical and Privacy Concerns:
   - Data Privacy: AI in education involves collecting and analyzing extensive student data, raising concerns about privacy and misuse. Chapter 5 - Data Privacy and Security Best Practices
   - Bias and Discrimination: AI systems can inherit biases from training data, leading to discriminatory outcomes against student groups. Chapter 2a - Ethical AI Integration (section on addressing bias and fairness)
2. Technical Challenges:
   - Reliability and Accuracy: AI systems may not always be reliable or accurate, impacting educational decisions. Reliability and Accuracy: Chapter 7 - Vetting and Implementing AI Tools/Vendors
   - Integrating AI into education systems can be costly and challenging. Integration Challenges: Chapter 7 - Vetting and Implementing AI Tools/Vendors
3. Pedagogical Concerns:
   - Personalization vs. Standardization: AI can offer personalized learning experiences but risks moving away from standardized curricula, affecting educational equity. Personalization vs Standardization: Chapter 4 - Balancing AI and Human Instruction
   - Balancing human educators and AI in education is complex and can impact teaching methods, authority, and student-teacher dynamics. Teacher/AI Roles: Chapter 4 - Balancing AI and Human Instruction
   - The impact of AI on long-term educational and social development is largely unknown. Long-term Outcomes: Chapter 3 - AI's Impact on Learning and Development
   - Dependency on Technology: Excessive reliance on AI may hinder critical thinking, creativity, and understanding. Dependency on Tech: Chapter 4 - Balancing AI and Human Instruction
   - Neglecting Broader Educational Goals: AI-driven education may prioritize market trends over critical thinking, civic responsibility, and artistic expression.
4. Accessibility and Equity:
   - Digital Divide: Unequal access to technology can widen the educational gap, leading to disparities in educational quality. Digital Divide: Chapter 8 - Ensuring Equitable Access to AI Education

5. Psychological and Cultural Impact:
   - Impact on Student Self-Perception: Continuous interaction with AI from a young age may affect how students perceive their intelligence, learning capabilities, and value. Student Self-Perception: Chapter 3 - AI's Impact on Learning and Development
   - Global AI solutions may erode local cultural, historical, and ethical values. Cultural Erosion: Chapter 6 - Localizing AI for Cultural Relevance
6. Legal and Policy Challenges:
   - Regulatory Compliance: Schools must stay updated with the evolving legal landscape around AI in education. Regulatory Compliance: Chapter 9 - Policy and Governance Frameworks
   - Accountability for AI failures or harm in education is complex. Accountability: Chapter 9 - Policy and Governance Frameworks

In the following chapters, we will explore these challenges in greater depth, focusing on practical strategies and frameworks for mitigating risks and harnessing the potential of AI to enhance education equitably and responsibly.

CHEATSHEET FOR TEACHERS:

Teacher Actions	Ethical & Privacy	Technical	Pedagogical	Accessibility & Equity,	Psychological & Cultural	Legal & Policy
Use an encrypted LMS and implement two-factor authentication.	✓					
Stay informed about data handling by subscribing to edtech newsletters and attending AI workshops.	✓					✓
Evaluate the diversity of AI training data. Provide feedback to developers and review AI tool performance.	✓		✓			
Learn about AI limitations & biases and independently verify AI outputs.	✓	✓	✓			
Collaborate with IT on AI needs by forming a working group and piloting small projects.		✓	✓			
Balance personalized learning and curriculum by using AI for teaching, support, and assessments.			✓			
Align AI lessons with curriculum goals. Automate feedback tasks and support diverse learning styles.			✓			
Lead discussions on AI technology using journaling and forums.			✓		✓	
Implement project-based learning that includes local content. Challenge AI outputs through critical thinking.			✓		✓	
Prepare non-digital materials and offer flexible submission options.				✓		
Encourage open conversations about AI impact. Foster critical thinking and media literacy.			✓		✓	
Emphasize human connection. Support the exploration of cultural identities. Collaborate with parents on AI impact.					✓	

CHEATSHEET FOR ADMINISTRATORS

Administrator Actions	Infrastructure & Access	Privacy & Data Security	AI Integration & Training	Curriculum Development	Ethical & Fair Use of AI	Policy & Compliance
Improve infrastructure: lend devices, enhance WiFi, include mobile-friendly content	✓			✓		
Ensure data privacy and security: conduct workshops, develop guidelines, audit practices		✓			✓	
Implement AI responsibly: address biases, develop roadmap, establish review processes			✓		✓	✓
Provide AI training and support: train staff, promote professional development, foster supportive environment			✓	✓	✓	
Enhance curriculum and teaching methods: implement data analytics, organize professional development, incorporate ethical AI foundations, develop balanced curriculum			✓	✓	✓	
Establish policies and accountability: organize compliance training, develop policies, set up reporting system						✓

CHEATSHEET FOR PARENTS

Parent Actions	Privacy & Data Security	Engagement & Advocacy	Educational Equity	Technological Access	Psychological & Cultural Impact	Policy & Compliance
Review and understand privacy policies of AI tools	✓					
Request demos of AI tools and communicate with teachers about AI		✓				
Advocate for stronger privacy measures, transparency systems, and AI integration		✓	✓			✓
Stay informed about AI integration and attend school meetings on technology		✓				
Initiate discussions about AI during parent-teacher meetings		✓				
Stay engaged in your child's education and AI's impact		✓	✓			
Participate in school board meetings and advocate for balanced AI use		✓	✓			✓
Support human interaction and AI balance, and engage in AI integration discussions				✓	✓	
Communicate with school officials about technology needs and utilize local resources for technology				✓		
Regularly discuss AI's impact on learning with your child and schedule check-ins about AI education					✓	
Advocate for creative and critical thinking in curriculum, and include local cultural content			✓		✓	
Form a parent group for cultural workshops and encourage diverse skills and perspectives at home					✓	
Attend school board meetings to inquire about AI policy compliance and research AI education laws. Ask teachers and administrators about AI tools and participate in school meetings on technology		✓				✓

2

ETHICAL AI INTEGRATION

Let's kick things off with a chat about Stuart Russell, a computer science professor at the University of California, Berkeley. He offers intriguing insights into the impact of AI on education. He's all about the "dance of teaching," saying that no matter how fancy our tech gets, there's nothing quite like a human teacher. He points out that, sure, AI is smart, but it's not great at getting the full picture, which is where teachers shine. As we see more AI in classrooms, Russell reminds us that we can't do without teachers' wisdom, especially when it comes to guiding and inspiring our students.

But here's the kicker: using AI in schools isn't about taking the reins away from our teachers. Instead, it's about making their lives a bit easier and helping them do what they do best, but even better. Teachers are the ones who really get what their students need. They come up with awesome teaching strategies and guide their students' ethical and moral growth. AI's role? It's there to back them up, keeping the real and human elements of education front and center.

According to Russell, *AI is still figuring out how to nail the teaching game.* It's pretty good at sharing information and grading tests. Still, it struggles with getting the nuances of teaching, like understanding students' feelings, tailoring lessons individually, and creating that warm, engaging learning vibe we all love. This just goes to show how irreplaceable our teachers are and why AI should play more of a support role instead.

MAKING IT REAL: AI IN ACTION

Let's examine how this plays out with Ms. Thompson and Mr. Garcia, two fictional but relatable teachers.

Ms. Thompson and the Adaptive Learning Platform

Ms. Thompson has been teaching high school biology for years. She brings an AI-powered learning platform that customizes itself to fit each student's learning pace.

Ms. Thompson uses the platform's insights to spot where her students need more help and tailors her lessons to meet those needs. It's a perfect example of how teachers can use AI as a tool to create personalized learning experiences without losing their touch.

Mr. Garcia and the AI Tutoring System

Then there's Mr. Garcia, our math teacher. He's all about using an AI tutoring system to spice up his lessons for students in various grades. This AI tool adjusts its challenges and feedback for each kid, making practice time super effective. Mr. Garcia blends this AI magic with his traditional teaching, keeping things fresh with group projects and class activities. It's a great showcase of how AI can enrich learning in ways that respect and build on a teacher's role.

By framing AI as an ally rather than a replacement, we're looking at a future where education is more personalized, efficient, and fun. Teachers still steer the ship, but now, they have some cool tools to help navigate the journey.

SIDEBAR EXERCISE: Prompt Comparison: Experience AI's Potential

Let's consider these two prompts for an AI language model:

Prompt A: "Summarize the key events of World War II."

Prompt B: "Provide an objective summary of World War II's major events, causes, and consequences from multiple perspectives."

Jot down 2 - 3 observations of how you expect the AI's responses to differ based on that nuanced phrasing.

Now, let's examine the AI's actual outputs…

Prompt 1 Output:

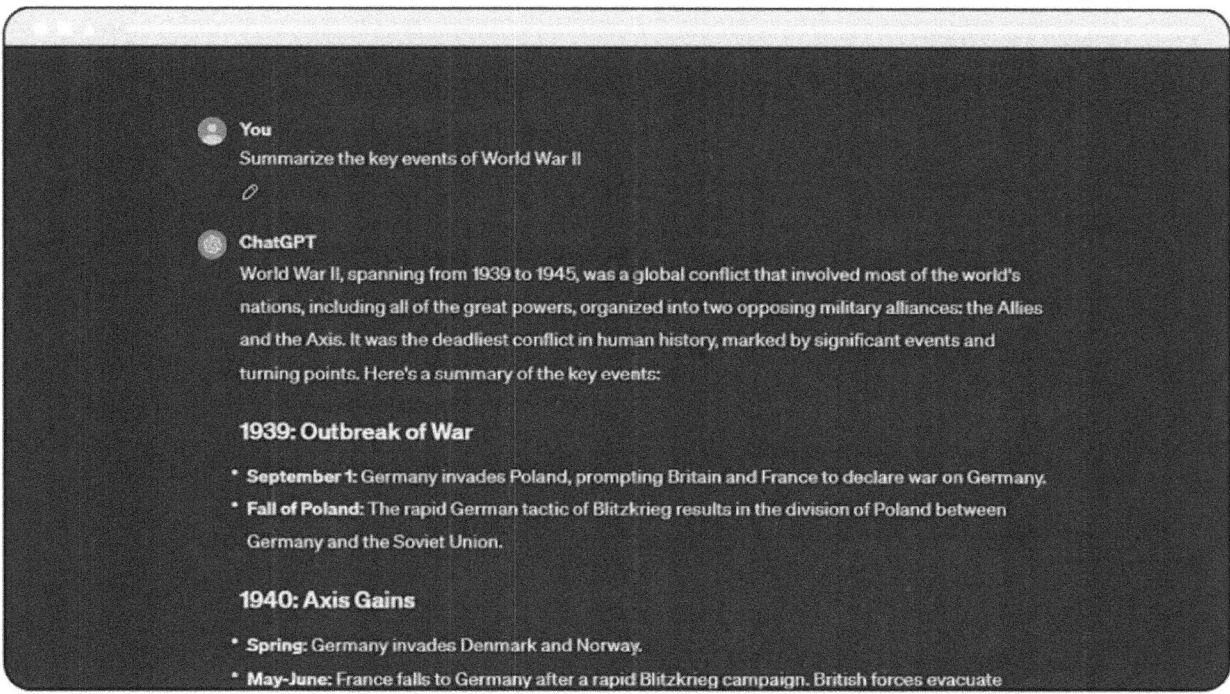

Prompt 2 Output:

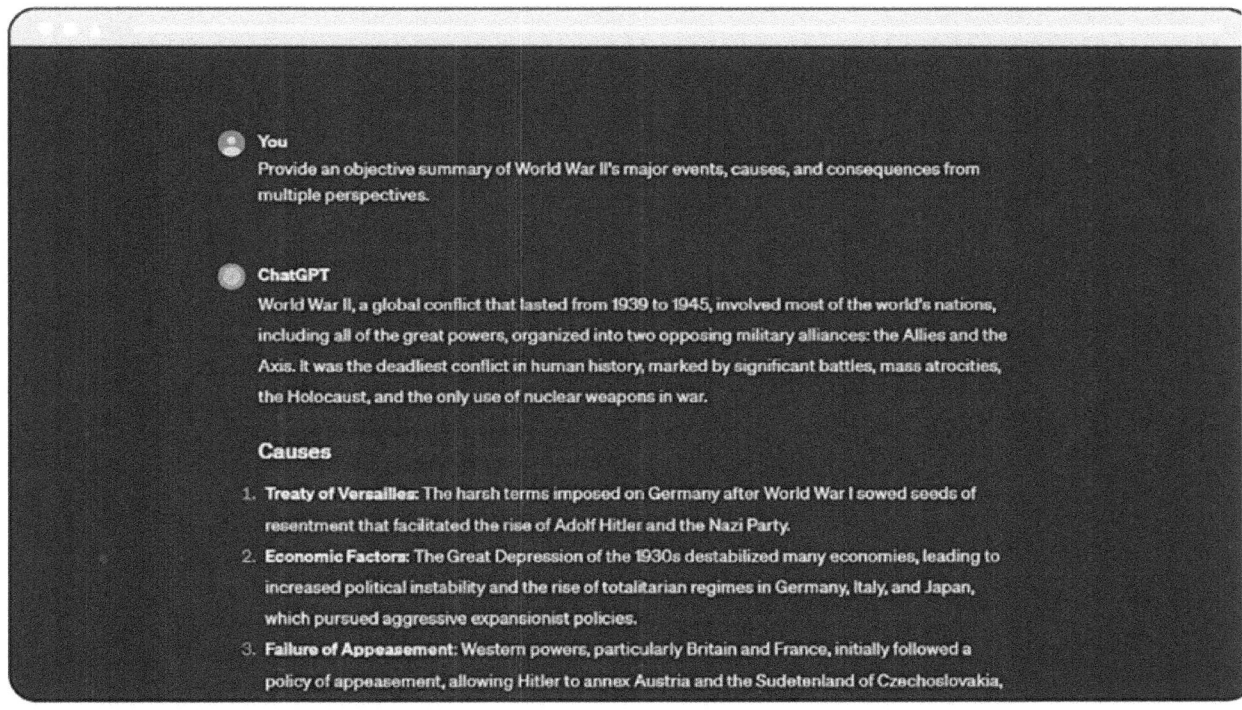

Why the difference?

The AI's responses adapt based on the nuances and framing of your prompts. As the NLP sidebar explains in Chapter 1, the language model recognizes patterns across vast datasets, allowing it to adjust outputs according to the provided context cues. Tweaking with specifics like requesting "objective, multi-perspective summaries" guides the AI to tailor its response style, scope, and content alignment.

As an educator, you unlock AI's collaborative potential through your prompting abilities. You can steer the AI to generate outputs that complement your instructional goals by crafting prompts that convey your desired objectives and parameters.

You have the power to shape AI's capabilities through intentional prompting. Use this to explore how AI can empower your classroom, support student growth, and amplify your pedagogical impact. The path forward is through curiosity, experimentation, and finding unique approaches that best harness its strengths for your learners.

SIDEBAR ENDS.

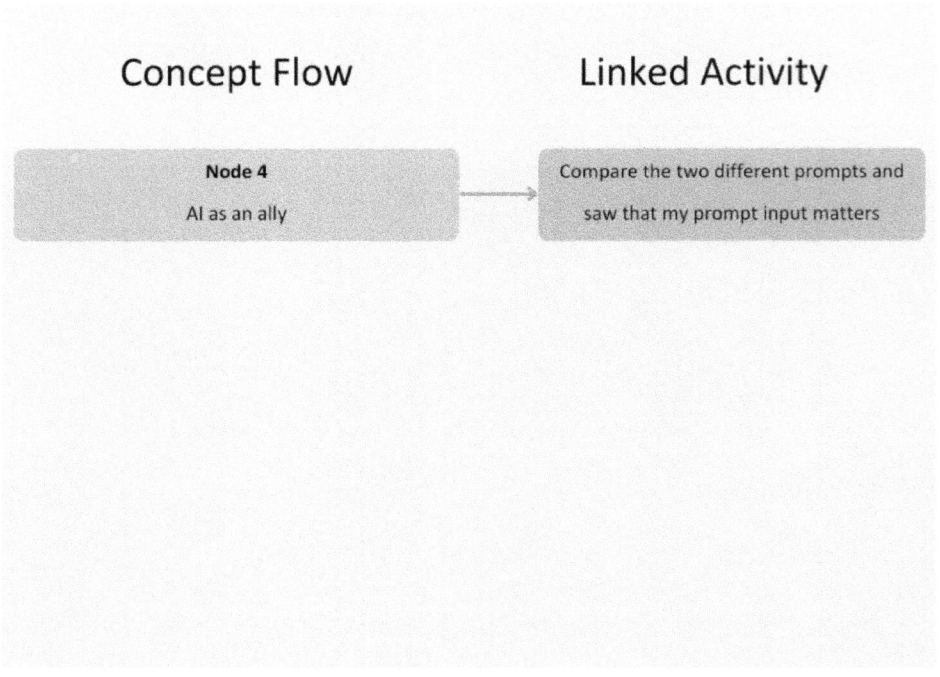

A JOURNEY THROUGH TIME: UNDERSTANDING COPYRIGHT'S ROOTS

Educators find themselves navigating uncharted territory regarding creativity and intellectual property in the classroom. Integrating AI as a tool for expression presents exciting opportunities and complex challenges. To ensure that we harness the power of AI

ETHICAL AI INTEGRATION | 37

responsibly, we must carefully consider how to uphold the fundamental principles behind copyright law, incentivizing innovation through the balanced protection of original ideas.

At the core of this issue lies the question of authorship when AI is involved in the creative process. If a student uses an AI writing assistant to generate content, who truly "owns" that work? The lines become blurred, and traditional notions of intellectual property rights are challenged. To find guidance, we can look to the philosophical origins of copyright law, such as the Statute of Anne, which sought to balance the open sharing of knowledge with the safeguarding of creators' rights.

The solution likely lies in finding a middle ground, guidelines that provide ethical guardrails for using generative AI as an inspirational aid, not a full substitute for personal expression. This could involve policies around prompting AI models, respecting intellectual property, and positioning AI as a supplemental creativity tool rather than replacing human ingenuity.

Ultimately, we want to nurture students' self-expression and original thinking while benefiting from AI's capabilities as a force multiplier. With a proactive, principles-based approach, we can harness AI's potential without undermining the incentives that have fueled human creativity throughout history.

SIDEBAR: The Statute of Anne: Balancing Innovation and Protection

The Statute of Anne, enacted in Great Britain in 1710, is widely regarded as the foundation of modern copyright law. This groundbreaking legislation introduced the concept of granting exclusive publishing rights to authors for a limited period.

The law struck a delicate balance between incentivizing creativity and promoting the dissemination of knowledge. It recognized that providing authors with temporary control over their works would encourage intellectual and artistic output, ultimately benefiting society as a whole.

The principle of balancing the interests of creators with those of the public lies at the heart of the Statute of Anne. The time-limited nature of the exclusive rights granted to authors motivated them to continue innovating while also ensuring that they would not persist indefinitely, which could hinder the free exchange of ideas. This philosophical underpinning, rewarding ingenuity through temporary protection, has shaped copyright frameworks worldwide for centuries.

However, the advent of the digital age has ignited debates about the extent to which modern copyright protection measures, such as Digital Rights Management (DRM), may be overreaching. DRM technologies restrict how digital content can be accessed, copied, and shared, even when such use is legal under fair use provisions. Critics argue that these controls undermine the original intent of copyright law, which sought to balance the incentives for creators with the broader societal benefits of access to knowledge and culture.

Furthermore, the Digital Millennium Copyright Act (DMCA) was enacted in the United States in 1998. It has faced criticism for its provisions that criminalize the circumvention of DRM, even in instances where the use of copyrighted material would not infringe upon the creator's rights. As we navigate new aspects of AI-generated content and its implications for authorship and intellectual property, it is crucial to reassess whether restrictive measures like DRM and the DMCA align with the goal of fostering innovation that copyright law originally aimed to promote.

The Statute of Anne reminds us of the importance of maintaining a balanced approach to copyright protection. This approach encourages creativity through reasonable safeguards while avoiding excessive restrictions that could stifle the innovation it seeks to nurture. As we adapt to the challenges posed by AI in the realm of authorship and creativity, we must strive to uphold the spirit of the Statute of Anne. This will ensure that copyright law remains a catalyst for progress rather than a hindrance to the free flow of ideas.

SIDEBAR ENDS

WHAT WILL MS. RAMIREZ DO?

Ms. Ramirez tapped her stylus against her tablet, brow furrowed as she reviewed the lesson plan for her high school art class. The school had just invested in a new AI art generation tool, and she was intrigued and apprehensive about integrating it. On the one hand, she could see the creative potential of having students craft unique prompts to spark the AI's imagination, which could lead to wonderfully unexpected artwork. It could be an inspiring way to blend technology with self-expression. However, Ms. Ramirez was also worried about authorship and copyright issues. **Was there a risk of inadvertent infringement if the AI model was trained on existing artwork?** And would relying too heavily on AI undermine her students' sense of personal creativity and ownership over their pieces?

As educators, the internal conflict Ms. Ramirez experienced likely resonates with many of us when considering AI art tools for the classroom. The questions around authorship, copyright implications of an AI's training data, and potential impacts on student creativity are complex ones without simple answers.

SIDEBAR: AI Art Generation Activity for Educators:

In this 25-35 minute activity, you'll have the chance to directly explore generative AI art tools and grapple with the implications around authorship and intellectual property rights. As you work through the prompting process yourself, consider how you might guide students through a similar exercise to build their understanding.

1. Open your device's generative AI art app like DALL-E, Stable Diffusion, or Midjourney.
2. Review the background on generative AI, authorship, and intellectual property rights:
   - AI models are trained on vast datasets to generate novel content.
   - This challenges traditional notions of human authorship and creativity.
   - There are ambiguities around who holds rights to AI-generated works.
3. Spend 5-10 minutes crafting text prompts to generate desired artwork. Experiment with different prompt styles and specificity.
4. Generate 3-5 AI artworks based on your prompts (5 mins).
5. Reflect on the following guided questions (5-10 mins):
   - To what extent did you exercise creative expression and agency through your prompts?
   - Should an AI's training data and process be considered part of its "creative" output?
   - How might we need to update intellectual property frameworks to incentivize AI progress while protecting human creators ethically?
6. Try prompting the AI to generate images in iconic copyrighted artistic styles (5 mins). Observe whether you could easily replicate well-known artworks. Consider if the app has sufficient safeguards against this, noting that some generators currently have more lax content filters than others.

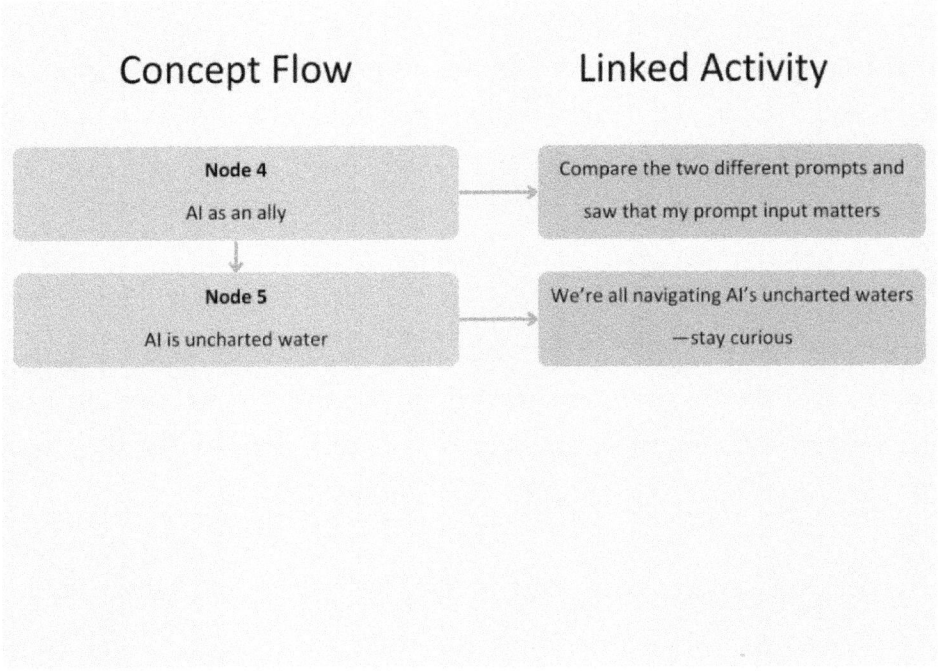

DEFINITION AND EMERGENCE OF AI IN CREATIVE FIELDS

The integration of AI into creative fields signifies an important shift in how we perceive and nurture human creativity. As we saw earlier, with our AI Art Generation activity, AI algorithms can now generate content that blurs the line between human and machine-made works. This prompts us to examine the implications, such as defining legal frameworks around copyright, shaping ethical boundaries, and developing educational initiatives for digital literacy.

Distinguishing AI-generated from human-created content is crucial for these reasons. We must deeply explore the mechanics behind AI's generative capabilities to understand the ramifications across legal, ethical, and educational domains. Our exploration of AI-generated content goes beyond appreciating the technology; it requires engaging with how AI will impact creative expression, ownership attribution and instilling ethical principles in future creators and innovators.

AT A CROSSROADS:

This book primarily aims to empower educators to navigate the responsible integration of AI into their teaching practices. However, educators also play a key role in guiding students on the appropriate use of AI tools for their learning. We are essentially dealing with two parallel decision frameworks here.

To provide clarity, I have created two distinct sections or "nodes" in this chapter:

1. When is it ethical and appropriate for an educator to utilize AI tools in their professional duties (lesson planning, grading, administrative tasks, etc.)?
2. When should students be permitted or encouraged to use AI tools to support their learning, and what guidelines should educators provide?

By separating these two perspectives, we can dive deeper into the nuances and considerations unique to each group's use of AI. For educators, we may explore factors like transparency, intellectual property, pedagogical implications, and modeling ethical practices. For students, the focus could be on developing critical thinking, avoiding academic dishonesty, accessibility/equity, and aligning with learning objectives.

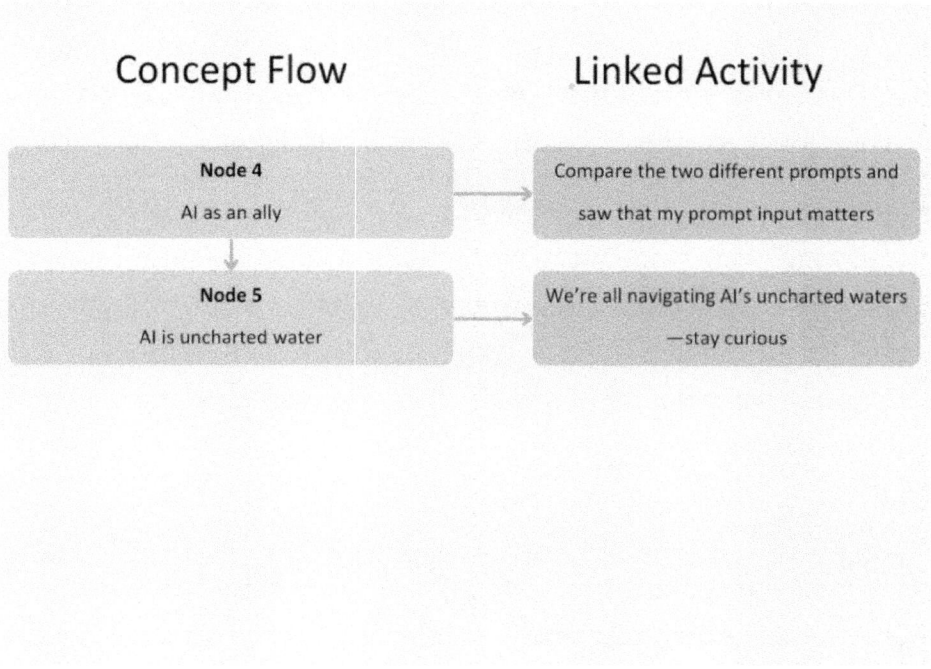

Before we proceed

While the ethical implications of AI are complex and worthy of deeper exploration, our aim is to equip you with actionable strategies for responsibly adopting these technologies. Instead of getting bogged down in theoretical debates, we will focus on real-world examples, straightforward best practices, and tangible steps to uphold principles like transparency in an accessible way. For this chapter, we will set aside the more abstract debates to prioritize practical guidance going forward. This is not meant to treat the topic as flippant but with the understanding that these issues deserve further discussion beyond the scope of this book. We also encourage you to sign up for our newsletter to stay informed on the latest developments and join the ongoing conversation around AI ethics in education.

COPYRIGHT IN THE DIGITAL CLASSROOM: BALANCING ACCESS AND OWNERSHIP WITH "FAIR USE"

"Fair use" plays a central role in the balance between access and ownership, allowing for the use of copyrighted materials for teaching and learning.

For educators, the fair use doctrine allows using copyrighted works for teaching purposes within specific limitations. However, interpretations can vary across borders, so educators must understand local and international copyright distinctions. This ensures their digital content's use aligns with relevant norms and laws. Additionally, as AI integration increases, data privacy and protection concerns become paramount. Regulations like the GDPR set high standards for handling personal data in educational contexts, requiring a balance between innovation and safeguarding individual rights.

From the student perspective, fair use provides some flexibility in accessing copyrighted materials for learning. Yet, students must be guided on respecting intellectual property and the ethical parameters around repurposing digital content. Moreover, data privacy literacy is crucial as they create their own digital works using AI tools. Students need to understand how these technologies may collect and use their personal information.

Transparency

The core principle of transparency regarding AI usage is simple: be upfront about it. Just like we would cite any other resource or tool used in our lessons, we should openly acknowledge when AI lent a hand. It can be as straightforward as saying, "I utilized an AI writing assistant to help draft this assignment prompt," or "The imagery for today's presentation was partially generated using AI art tools."

The key is normalizing AI as just another innovative technology in our educational toolkit, not something to be feared or obscured. We're not asking students to solve complex moral dilemmas, just to exercise the same honesty and integrity we've always expected.

In fact, being transparent about AI can be an opportunity to model ethical behavior and have productive discussions around authorship, intellectual property, and appropriately crediting sources. We can share our own experiences vetting AI tools, setting boundaries for acceptable use, and navigating potential pitfalls.

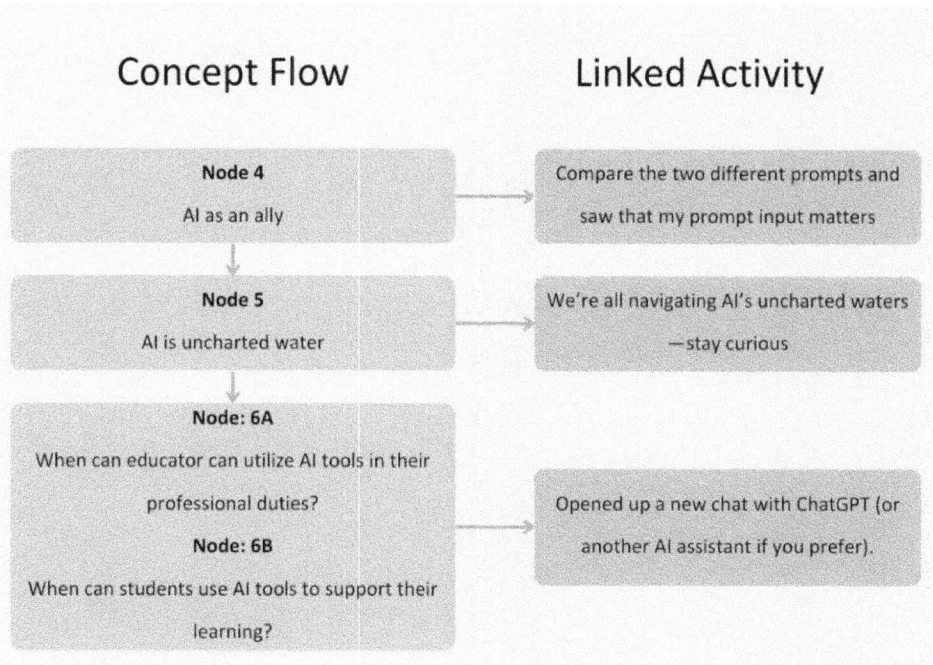

SUGGESTED FRAMEWORK

We recognize the immense potential for these tools to augment teaching practices and provide more personalized, engaging learning experiences. However, we must be thoughtful and principled in harnessing their power in education.

This framework aims to balance empowering teachers while maintaining the importance of human-led instruction. Like balancing on one leg, it requires constant adjustment to maintain equilibrium. As AI capabilities evolve, our approach to integrating them responsibly must adapt as well.

While not an official policy mandate, these guidelines have been benchmarked against existing ethics principles and district policies. They provide a robust starting point that can be adapted based on your specific context and needs.

Our central viewpoint of prohibiting the use of unvetted, "raw" output in any educational materials or assessments is non-negotiable. Presenting AI-generated content without human oversight and curation contradicts established pedagogical best practices and could undermine learning outcomes.

We encourage you to review this framework through the lens of your educational philosophies and values. Ultimately, our goal is to uphold the primacy of the teacher-student relationship and ensure it remains a supplemental tool that extends your capabilities as an educator, not a replacement for your professional expertise.

Guiding Principle: No Raw AI Output

Educators should use AI-generated content only after thoroughly reviewing, editing, and customizing the output to ensure accuracy, alignment with learning objectives, and the absence of bias or inappropriate content. Raw, unvetted AI output should never be presented to students in any educational context.

Level 0 (Red) - No AI Use Permitted

- Applicable to: Summative assessments, high-stakes evaluation of student work, official student records/reports
- Acceptable Use: All content must be 100% human-authored with no AI involvement.
- Unacceptable Use: Using AI to generate any portion of test questions, report card comments, or other formal assessment/evaluation materials.

Level 1 (Yellow) - AI for Research/Reference and Ideation Only

- Applicable to: Gathering information and resources to inform lesson planning; brainstorming initial instructional ideas
- Acceptable Use: Leveraging AI tools to supplement human-led lesson development; using AI to generate rough outlines or starting points that will require substantial teacher customization.
- Unacceptable Use: Directly incorporating AI-generated lesson plans, curricula, activities, or other core instructional content without significant editing to align with student needs and learning standards.

Level 2 (Green) - AI as a Support Tool

- Applicable to: Creating supplementary teaching materials; providing formative feedback on student work

- Acceptable Use: Using AI to generate drafts of worksheets, activities, visual aids, or feedback comments that are then heavily revised and personalized by the teacher and clearly disclosing when AI assistance was involved.
- Unacceptable Use: Presenting any AI-generated educational content or student feedback to students without thorough human review, refinement, and acknowledgment.

Level 3 (Green) - AI as Teaching Assistant

- Applicable to: Implementing AI tools to provide additional support, differentiation, and interactivity during instruction
- Acceptable Use: Vetting and testing AI-powered tutoring systems, adaptive learning platforms, etc., to augment teacher-led lessons, providing full transparency to students/parents about the tools used and all data privacy measures in place.
- Unacceptable Use: Relying on AI tools to autonomously deliver core instruction or utilizing unapproved platforms without properly vetting their efficacy, accuracy, security, and alignment with district policies.

Sidebar: The Importance of Human Oversight

Consider this example of an AI-generating test question on cellular respiration for a high school biology exam:

Raw AI Output:

"What is the process by which cells convert nutrients into ATP?

A. Photosynthesis
B. Cellular respiration
C. Fermentation
D. Mitosis

In cellular respiration, what is the final electron acceptor?

A. Carbon dioxide
B. Water
C. Oxygen
D. Glucose

While technically providing biology questions on the topic, there are several potential issues with using this unedited AI output:

- Questions may not align with specific learning standards
- Questions could be ambiguous, misleading, or factually inaccurate
- No distractors or explanations to make the questions instructionally valuable
- Language may not be appropriate for the student level

Now, let's see those same questions after review and revision by the human teacher:

"Cellular respiration is a process that converts nutrients into ATP, the energy currency of cells. Which accurately describes this process?

A. Combines CO_2 and H_2O using sunlight to produce glucose
B. Breaks down glucose through oxidation reactions, with O_2 as the final electron acceptor
C. Converts ADP into ATP through fermentation without oxygen
D. The process where cells divide to produce new cells

Explanation: Cellular respiration breaks down glucose to produce ATP for energy. In the electron transport chain, oxygen acts as the final electron acceptor, differentiating it from photosynthesis, fermentation, and mitosis."

In the revised version:

- Questions are clearly worded for the level.
- Distractors test common misconceptions.
- An explanation reinforces key concepts.

This highlights why prohibiting raw, unvetted AI output is crucial. Human expertise is required to ensure accuracy, instructional value, and alignment with learning goals.

SIDEBAR ENDS

Sidebar: Bias in AI algorithms

Bias in AI algorithms can originate from various sources and often reflects the prejudices in the training data. AI algorithms learn any existing inequalities, stereotypes, or biases from historical data. For instance, if an AI system is trained on job application data predominantly featuring a specific gender or ethnic group in certain roles, it may develop a bias, favoring those groups for similar jobs. This bias can be explicit, using discriminatory variables in decision-making, or implicit, where data correlations lead to biased outcomes.

Addressing bias in AI requires a multifaceted approach, including de-biasing training data, using algorithms to mitigate bias, and auditing AI systems for biased outcomes. However, technical solutions alone are insufficient to address bias in AI. Empowering students' digital literacy through AI plays a crucial role.

Empowering students with digital literacy in the context of AI involves educating them on how AI technologies work and their ethical, social, and cultural implications. It includes critical thinking, data literacy, understanding algorithmic decision-making processes, and recognizing biases in AI systems. This enhances students' ability to evaluate AI technologies, understand their limitations, and advocate for ethical AI applications.

Promoting digital literacy involves integrating AI education into school curriculums, hosting ethical AI workshops and seminars, and creating platforms for discussing AI's social impact. These efforts should aim to cultivate informed AI users and developers who are mindful of ethical considerations and potential bias.

3

DIGITAL LITERACY IN AN AI-ENHANCED EDUCATION

Digital literacy is paramount for students to thrive. Basic computer skills are no longer enough, the future demands harnessing AI's power, navigating big data, and thinking creatively. By 2027, the most sought-after skills will be analytical thinking, creative thinking, and the ability to use AI and big data. As educators, we must foster these digital literacy skills to empower students to survive and confidently navigate an AI-centric world.

Digital literacy goes beyond employability; it's about equipping students with the tools to critically evaluate the AI systems shaping their lives. Students need ethical awareness to thoughtfully engage with AI, from social media algorithms to automated decision-making.

We'll provide a roadmap for integrating these competencies across curricula. We'll explore innovative approaches like the "flipped desk" model that puts AI tools directly into students' hands.

Throughout the chapter, we'll build towards a powerful culminating experience: the AI hackathon. These project-based learning events challenge students to apply their skills to real-world problems, fostering collaboration, creativity, and ethical reasoning.

By the end of this chapter, you'll have a toolkit for weaving AI literacy into your classroom and a step-by-step guide to hosting your own hackathons.

WHAT IS DIGITAL LITERACY?

Traditional digital literacy has been defined as the ability to effectively use digital technologies, communication tools, and networks to locate, evaluate, use, and create information.

In other words, digital literacy refers to the skills and knowledge we need to thrive in the digital world.

However, in the age of AI, this definition needs to be expanded to encompass AI-specific skills and competencies.

As a subset of digital literacy, AI literacy involves critically evaluating AI systems, understanding their underlying algorithms and data sources, and recognizing potential biases or limitations.

Expanding the definition of digital literacy to include AI-specific skills and competencies involves several key elements:

1. **Understanding AI fundamentals:** A basic understanding of AI concepts, such as machine learning, neural networks, and natural language processing, is essential for navigating the AI-driven world. (Discussed in Chapter 1)
2. **Ethical considerations:** Recognizing the ethical implications of AI systems, including issues related to privacy, bias, and transparency, and being able to make informed decisions about their use. (Discussed in Chapter 2)
3. **Critical evaluation:** Developing the ability to critically evaluate AI systems, question their outputs, and understand their potential limitations and biases.
4. **Data literacy:** Understanding the importance of data quality, data privacy, and data management in the context of AI systems.
5. **Computational thinking:** It is crucial to develop problem-solving skills and break down complex problems into smaller, more manageable parts when working with AI systems.

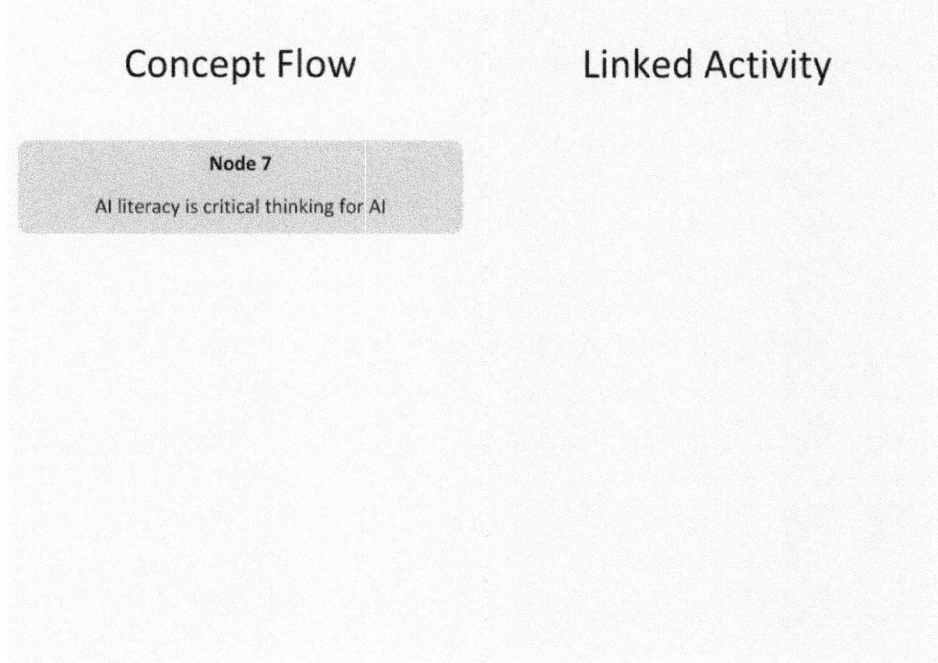

ESSENTIAL SKILLS FOR EDUCATORS AND STUDENTS

Meet Ms. Dataphobe, a well-meaning but data-illiterate teacher who uses AI to analyze students' performance data. Excited about the prospect of data-driven insights, she eagerly feeds the data into the AI system without adequately checking its quality or representativeness.

She missed the fact that the data was riddled with errors and biases. Some students accidentally entered their ages instead of their grades, while others used emojis to express their feelings about the subject matter. The AI, unable to distinguish between meaningful and erroneous data, generates wild and nonsensical insights.

Like a fortune teller reading tea leaves, the AI predicts that students who used pizza emojis will become master chefs, while those who used their age instead of grades are flagged as potential failures.

We might laugh at Ms. Dataphobe, but we know how easy it would be for students to enter incorrect data, leading to flawed AI predictions.

Data Literacy

Data literacy is the critical skill of collecting, managing, evaluating, and applying data. In the age of AI, it's crucial for educators and students.

As we have seen, the quality and representativeness of data directly impact an AI system's performance and decision-making capabilities.

Data literacy skills are vital for educators to effectively teach data-related concepts, analyze student performance data, and make informed decisions about curriculum and instruction.

On the other hand, students need data literacy to evaluate information sources critically, interpret data visualizations, and communicate findings effectively when working with data-driven projects or assignments. **Data literacy also helps students understand the implications of sharing personal data online and how AI systems can use it.**

Practical examples of data literacy in action include:

- Analyzing datasets to identify potential biases or errors before using them to train an AI model
- Interpreting data visualizations in news articles or research reports to understand underlying trends and patterns
- Collecting and managing data from various sources (e.g., surveys, sensors, online platforms) for school projects or research studies
- Evaluating the credibility and relevance of data sources when conducting research or making decisions.

Computational Thinking

Computational Thinking (CT) is a problem-solving approach that breaks down complex issues into smaller, manageable parts. It helps you recognize patterns and develop step-by-step solutions.

CT is great for improving logical reasoning, creativity, and critical thinking skills. For example, in education, CT can be applied in lots of different ways:

- Math: Encouraging students to break down complex equations, identify patterns, and create algorithms for efficient problem-solving
- Science: Designing systematic experiments, collecting and analyzing data, and drawing logical conclusions
- Language Arts: Structuring narratives, identifying themes, and systematically analyzing literary elements

CT is a mindset that goes beyond specific subjects or technologies. It helps students become better at solving problems and developing new ideas. If Ms. Dataphobe had used CT, she could have split the data analysis into smaller tasks, found and fixed issues with the data, and made a step-by-step plan to get reliable results.

Critical evaluation of AI systems

Critical evaluation of AI systems involves analyzing AI models' performance, limitations, and potential biases to ensure they are accurate, fair, and ethical. By understanding the underlying algorithms and data sources, we can identify issues and take corrective measures for responsible AI development and deployment.

The case of Microsoft's Tay chatbot, which quickly learned offensive language from online interactions, highlights the importance of robust evaluation before real-world deployment. Proper testing and safeguards could have prevented its inappropriate behavior.

As educators, we must foster critical evaluation skills in our students. By teaching them to assess AI systems for transparency, accountability, and alignment with societal values, we empower them to become responsible creators and consumers of AI.

FLIP THE DESK: INTEGRATING DIGITAL LITERACY ACROSS CURRICULA

Just as the "flipping the classroom" model upends the traditional approach of lecturing first and applying knowledge later, we suggest a tongue-in-cheek term, "flip the desk," regarding AI in education. Instead of viewing AI as an obstacle to overcome, we should embrace it as an inevitable part of our students' future. By putting AI tools directly into their hands through a "flipped desk" approach, we shift the power dynamic and empower students to take an active role in their AI literacy journey. This hands-on, experiential learning encourages critical engagement, where students leverage AI capabilities for assignments and projects but also analyze the outputs through a critical lens.

This "flipping the desk" approach encourages experiential, hands-on learning where students critically engage with AI through assignments and projects. For example, they could use language models to generate content **but then analyze the output for potential biases, inaccuracies, or ethical concerns.** Or they could leverage AI tools to explore complex topics from multiple angles, solving real-world problems while exercising human judgment. The key is creating relevant, meaningful assignments that show the value of AI capabilities while reinforcing the importance of applying creativity, moral reasoning, and critical thinking skills. By acknowledging AI's inevitability in their future careers, we can develop guidelines around acceptable use, how to properly cite AI contributions, set boundaries, and make informed decisions about when this technology is appropriate. Ultimately, our role is not to construct barriers but to equip students with the critical AI literacy skills to navigate this responsibly and ethically. An open, transparent approach allows us to model best practices while empowering the next generation to harness AI's potential as a powerful tool.

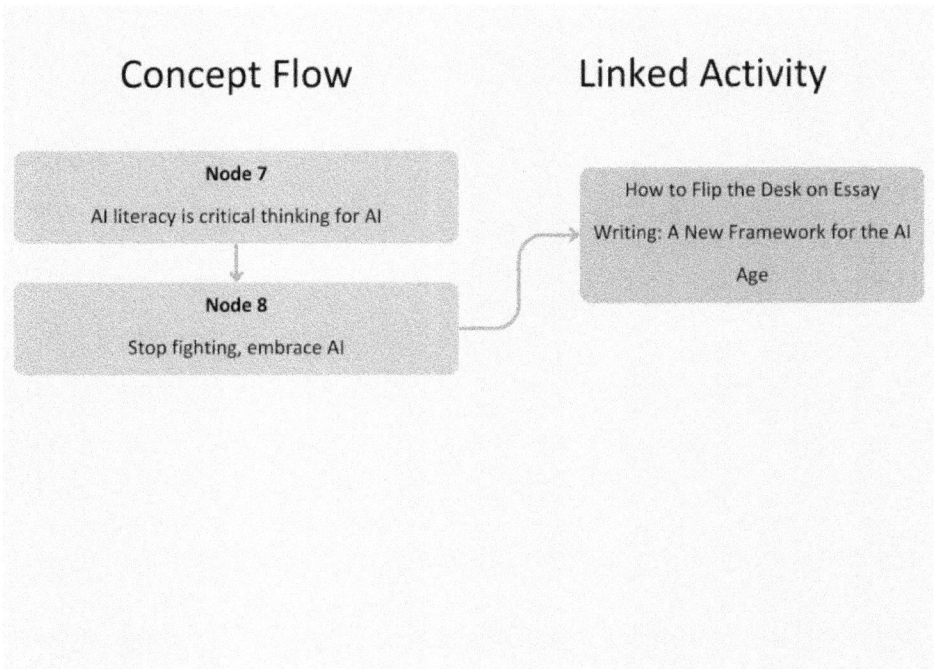

How to Flip the Desk on Essay Writing: A New Framework for the AI Age

Let's face it: it's easy for students to use AI to generate essays. As AI gets better at producing human-like text, we will find it increasingly difficult to distinguish AI-written essays from those written by actual students, challenging traditional assessment and evaluation methods. The essay, long a staple of academic writing, is now vulnerable to disruption by AI-powered tools.

It's time to "flip the desk" on the traditional essay assignment. Instead of viewing AI as a threat to academic integrity, we should embrace it as an opportunity to cultivate critical AI literacy skills. By putting these tools directly into students' hands and empowering them to take an active role in their learning journey, we can foster a new generation of digitally savvy, ethically grounded writers. Plus, we can make our lives easier.

The Flipped Desk Framework:

1. **AI as a Co-Pilot:** Encourage students to leverage AI throughout the writing process, from brainstorming to drafting to revision. Teach them to use AI-generated text as a starting point for their ideas, but emphasize the importance of substantial human refinement and editing.

2. **Critical AI Analysis:** Incorporate lessons on identifying the quirks and limitations of AI-generated writing. Have students critically evaluate AI outputs for logical inconsistencies, factual errors, or stylistic oddities, fostering their ability to discern when and how to use these tools effectively.

3. **Innovative Assessments:** Diversify writing assignments beyond the traditional essay format. Explore multimedia projects, debates, research reports, or other tasks that require students to gather, evaluate, and synthesize information from various sources while exercising human judgment.

4. **Transparent AI Usage:** Students must clearly cite any AI-generated text used in their work, along with the prompt provided to the AI. This transparency promotes accountability and allows educators to assess how effectively students integrate AI into their writing process.

5. **Collaborative AI Guidelines:** Involve students in developing honor code language around the acceptable use of AI in their work. Discuss AI writing detection tools transparently, explaining their role in upholding academic integrity while acknowledging their limitations.

6. **Metacognitive Reflection:** Have students maintain a log or journal documenting their use of AI throughout the writing process. Prompt them to reflect on how AI supports and enhances their writing and its limitations, fostering valuable insights into their own learning.

7. **Celebrating Human Skills:** Highlight the crucial role of creativity, nuance, empathy, and contextual understanding in effective writing. Use examples or case studies to illustrate the pitfalls of over-relying on AI without human oversight, reinforcing the importance of these uniquely human abilities.

HACKATHONS: THE ULTIMATE TOOL FOR TEACHING DIGITAL LITERACY IN AN AI-DRIVEN WORLD

The hackathon briefs in this chapter show how AI can be used in classrooms and other educational settings. They take a "flipped desk" approach where students become active creators, using AI to solve real-world problems. Through these hackathons, students:

- Explore AI capabilities and outputs
- Examine the ethical implications of AI technologies
- Gain insights into how technology shapes our world

As students' knowledge of AI's potential and limitations grows, they can start to recognize the importance of combining AI with human skills. So, use the below-suggested hackathons to increase the student's nuance, empathy, and creative problem-solving skills:

1. **AI Ethics Challenge:** Students work in teams to identify potential ethical issues with an existing or hypothetical AI application. They then propose solutions and safeguards to mitigate those concerns through product design, policies, or public education campaigns.
2. **AI for Good Hackathon:** The event will focus on developing AI prototypes to tackle real-world social or environmental challenges. Students will research problems, gather data, and build AI models to generate innovative solutions under mentorship from industry experts.
3. **AI Creativity Showcase:** Instead of coding, students use AI tools like language models, image generators, and music creation AIs to produce original creative works. Judging criteria could include creativity, technical skill in prompt engineering, and responsible AI use.

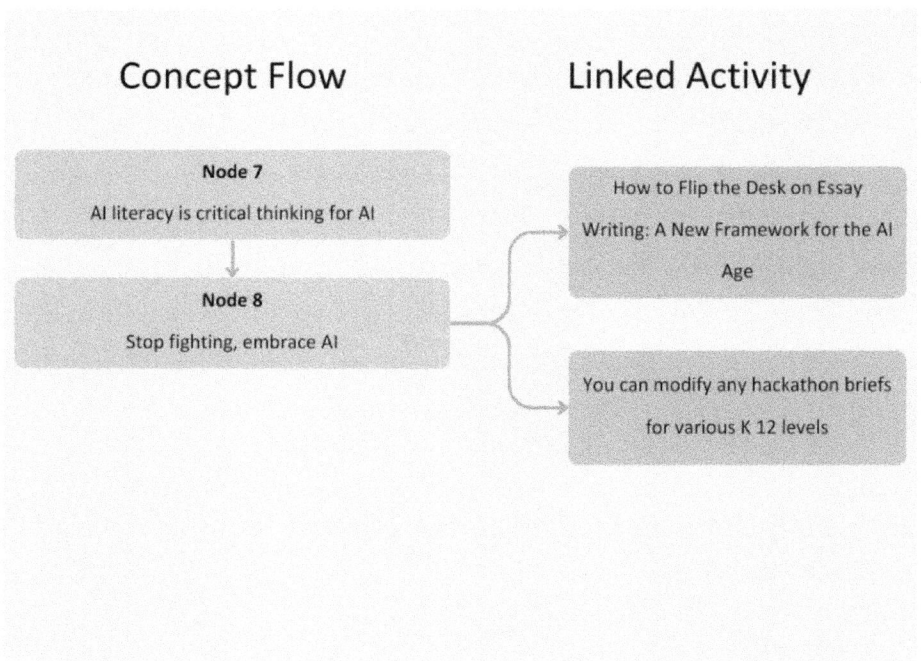

Sidebar

You can modify any hackathon briefs for various K 12 levels by adjusting the AI applications' complexity, research depth, and critical analysis using ChatGPT and the prompt below.

For younger children: Teachers can focus on more straightforward AI applications and provide extra guidance for younger children.

As students advance, teachers can increase complexity, demand more research, and prompt refined ethical analysis and solution formulation.

The included grading rubric can also be tailored to match each grade level's expectations.

How to change the brief: Use ChatGPT to input the project text and marking rubric, ask for customization suggestions for differing grade levels, and generate ideas for suitable AI applications, research resources, and evaluation criteria.

Suggest prompt: *I plan to use the [INSERT NAME OF HACKATHON HERE] with my [grade level] students. Could you provide suggestions on how to adapt the project, including:*

1. *Age-appropriate AI applications for students to study*
2. *Specific research resources suitable for this grade level*
3. *Adjustments to the project steps and expectations to match students' developmental stage*
4. *Modifications to the marking rubric to align with grade-level expectations: Please provide your recommendations on how I can effectively implement this project with my students.*

Be sure to include the briefs in the prompt below so that ChatGPT knows what to edit.

THE AI ETHICS CHALLENGE: A PROJECT-BASED LEARNING GUIDE

Objective: The AI Ethics Challenge is a project-based learning activity designed to engage students in critically analyzing AI applications, identifying potential ethical risks, and developing strategies to mitigate these concerns. By participating in this challenge, students will gain a deeper understanding of fundamental ethical principles in AI development and deployment, such as transparency, fairness, privacy, and accountability.

Materials:

- AI Ethics introduction presentation
- Research resources (online databases, articles, case studies)
- Proposal template
- Evaluation rubric

Step-by-Step Guide:

Introduction to AI Ethics (1 class session)

- Present key ethical principles in AI development and deployment.
- Discuss real-world examples of AI ethics issues.
- Introducing the AI Ethics Challenge project.

Team Formation (1 class session)

- Divide the class into small groups of 3-5 students.
- Ensure that each team has a diverse mix of backgrounds and perspectives.

AI Application Selection (1 class session)

- Teams choose an existing AI application (e.g., facial recognition, predictive policing, hiring algorithms) or develop a hypothetical AI use case.
- Encourage teams to consider applications relevant to their interests or current events.

Research Phase (2-3 class sessions)

- Teams investigate their chosen AI application.
- Identify potential ethical risks, biases, and societal impacts through research.
- Provide resources such as online databases, articles, and case studies.

Ideation (1-2 class sessions)

- Teams brainstorm solutions to mitigate identified ethical concerns.
- Consider product design changes, policies, public education campaigns, etc.
- Encourage creative problem-solving and out-of-the-box thinking.

Proposal Development (2-3 class sessions)

- Teams create a detailed proposal using the provided template.
- Outline the AI application, ethical issues identified, and proposed solutions.
- Emphasize clear, concise writing and well-structured arguments.

Presentations (1-2 class sessions)

- Teams present their proposals to the class.
- Allocate time for questions and feedback from peers and the educator.
- Foster a supportive, constructive learning environment.

Peer Evaluation (1 class session)

- Students provide a constructive critique of each other's proposals.
- Analyze strengths and areas for improvement using the provided evaluation rubric.
- Encourage respectful, actionable feedback.

Reflection (1 class session)

- Facilitate a class discussion reflecting on critical lessons learned.
- Discuss insights on responsible AI development and deployment.
- Encourage students to share personal growth and "aha" moments.

Iteration (1-2 class sessions)

- Provide time for teams to refine their proposals based on feedback.
- Encourage the incorporation of peer and educator suggestions.
- Collect final proposals for grading and feedback.

AI Ethics Challenge Marking Rubric

Depth of Research and Understanding of Ethical Issues (30 points)

- Exemplary (25-30): Thorough research demonstrating a deep understanding of the AI application and its potential ethical implications. Identifies and analyzes multiple relevant ethical issues.
- Proficient (20-24): Adequate research, demonstrating a solid understanding of the AI application and its key ethical implications. Identifies and analyzes some relevant ethical issues.
- Developing (15-19): Limited research demonstrating a basic understanding of the AI application and its ethical implications. Identifies and analyzes a few ethical issues but may lack depth.
- Beginning (0-14): Insufficient research, demonstrating a lack of understanding of the AI application and its ethical implications. Fails to identify or analyze relevant ethical issues.

Creativity and Feasibility of Proposed Solutions (25 points)

- Exemplary (21-25): Proposes innovative, practical, and well-thought-out solutions that address the identified ethical issues. Solutions are feasible and demonstrate creative problem-solving.
- Proficient (16-20): Proposes relevant and mostly feasible solutions that address the identified ethical issues. Solutions demonstrate some creativity and practicality.
- Developing (11-15): Proposes solutions that partially address the identified ethical issues. Solutions may lack creativity, feasibility, or depth.
- Beginning (0-10): Proposes solutions that fail to address the identified ethical issues or are impractical. Solutions lack creativity and feasibility.

Clarity and Organization of the Written Proposal (20 points)

- Exemplary (17-20): The proposal is well-organized, clear, and concise. Arguments are logically structured and easy to follow. Writing is free of errors and professionally presented.
- Proficient (13-16): The proposal is mostly well-organized and clear. Arguments are generally logical and easy to follow. Writing may contain minor errors but is overall professionally presented.

- Developing (9-12): The proposal is somewhat organized and clear. However, the arguments may be difficult to follow at times. The writing may contain several errors and lack professional presentation.
- Beginning (0-8): The proposal is poorly organized and unclear. Arguments are illogical or difficult to follow. Writing contains numerous errors and lacks professional presentation.

Effectiveness of the Presentation and Response to Questions (15 points)

- Exemplary (13-15): Presentation is engaging, informative, and well-prepared. Team members effectively communicate their ideas and respond to questions thoughtfully and accurately.
- Proficient (10-12): Presentation is informative and mostly well-prepared. Team members communicate their ideas adequately and respond to questions satisfactorily.
- Developing (7-9): The presentation is somewhat informative but may lack preparation. Team members communicate their ideas with some difficulty and may struggle to respond to questions.
- Beginning (0-6): Presentation is uninformative, unprepared, or fails to communicate ideas effectively. Team members are unable to respond to questions adequately.

Individual Contributions and Growth (10 points)

- Exemplary (9-10): Individual demonstrates significant contributions to the team project and personal growth through reflections and peer evaluations.
- Proficient (7-8): Individual demonstrates adequate contributions to the team project and some personal growth through reflections and peer evaluations.
- Developing (5-6): Individual demonstrates limited contributions to the team project and minimal personal growth through reflections and peer evaluations.
- Beginning (0-4): Individual demonstrates insufficient contributions to the team project and lacks evidence of personal growth through reflections and peer evaluations.

Total Points: 100

THE AI FOR GOOD HACKATHON: A PROJECT-BASED LEARNING GUIDE

Objective: The AI for Good Hackathon is a project-based learning activity designed to engage students in developing AI-driven solutions for real-world social or environmental challenges. By participating in this hackathon, students will gain hands-on experience applying AI skills to create tangible solutions for issues they care about. They will also develop ethical reasoning and an appreciation for AI's potential for social impact.

Materials:

- Problem identification and research resources
- Data collection and curation tools
- AI model development software and resources
- User testing and feedback collection tools
- Impact assessment frameworks
- Presentation templates and guidelines

Step-by-Step Guide:

Problem Identification (1-2 class sessions)

- Students research and identify pressing social or environmental issues they are passionate about (e.g., climate change, food insecurity, healthcare access).
- Encourage students to select issues with available data and potential for AI-driven solutions.

Data Collection (2-3 class sessions)

- Guide students in ethically sourcing and curating relevant data sets related to their chosen problem area.
- Teach best practices for data collection, cleaning, and preprocessing.

Team Formation (1 class session)

- Create cross-functional teams with diverse skills, coding, design, subject matter expertise, etc.
- Ensure that each team has a balance of technical and non-technical skills.

Problem Framing (1-2 class sessions)

- Teams clearly define the specific problem they aim to solve and their goals/metrics for success.
- Encourage teams to break down their problem into manageable sub-problems.

Ideation (1-2 class sessions)

- Facilitate brainstorming sessions on potential AI-driven solutions, leveraging techniques like design thinking.
- Encourage teams to consider multiple approaches and evaluate their feasibility.

Model Development (4-6 class sessions)

- With mentorship from educators and subject matter experts, teams build and iteratively refine AI models to address their problems using the data collected.
- Provide resources and guidance on model selection, training, and evaluation.

User Testing (1-2 class sessions)

- Teams gather feedback from potential users/stakeholders on their AI solution prototypes.
- Encourage teams to use the feedback to improve and refine their solutions.

Impact Assessment (1-2 class sessions)

- Teams evaluate the potential positive and negative implications their AI solution could have.
- Guide teams in considering ethical, social, and environmental factors.

Final Presentations (1-2 class sessions)

- Teams pitch their AI prototypes and findings to a panel of expert judges and receive feedback.
- Encourage teams to present their problem, solution, and potential impact clearly and persuasively.

Next Steps (1 class session)

- Top solutions are awarded prizes/funding to continue developing AI applications for real-world deployment.
- Encourage all teams to reflect on their learning and consider future opportunities to apply their skills for social good.

AI for Good Hackathon Marking Rubric

Relevance and importance of the Problem Addressed (20 points)

- Exemplary (17-20): The problem addressed is highly relevant, pressing, and has significant potential for social or environmental impact.
- Proficient (13-16): The problem addressed is relevant and has moderate potential for social or environmental impact.
- Developing (9-12): The problem addressed is somewhat relevant but may lack urgency or potential for significant impact.
- Beginning (0-8): The problem addressed is irrelevant or lacks potential for meaningful social or environmental impact.

Quality and Ethical Considerations of Data Collection and Usage (20 points)

- Exemplary (17-20): Data collected is high-quality, relevant, and ethically sourced. Data usage follows best practices and considers privacy and security.
- Proficient (13-16): Data collected is adequate and mostly relevant. Ethical considerations are present but may not be comprehensive.
- Developing (9-12): Data collected is limited or partially relevant. Ethical considerations are minimal or not well-addressed.
- Beginning (0-8): Data collected is insufficient, irrelevant, or raises significant ethical concerns.

Effectiveness and Creativity of the AI Solution Developed (25 points)

- Exemplary (21-25): The AI solution effectively addresses the identified problem, demonstrates creativity, and has strong potential for real-world impact.
- Proficient (16-20): The AI solution adequately addresses the identified problem and shows some creativity but may have limitations in real-world application.

- Developing (11-15): The AI solution partially addresses the identified problem but lacks creativity or has significant limitations in Effectiveness.
- Beginning (0-10): The AI solution fails to address the identified problem effectively and lacks creativity.

The Thoroughness of Impact Assessment and Consideration of Implications (20 points)

- Exemplary (17-20): The impact assessment is thorough, considering a wide range of potential positive and negative implications. Ethical, social, and environmental factors are well-addressed.
- Proficient (13-16): The impact assessment covers key implications but may not be comprehensive. Ethical, social, and environmental factors are considered but not deeply explored.
- Developing (9-12): The impact assessment is limited and may not consider all relevant implications. Ethical, social, and environmental factors are minimally addressed.
- Beginning (0-8): The impact assessment is insufficient or lacks consideration of important implications. Ethical, social, and environmental factors are not well-addressed.

Clarity and Persuasiveness of the Final Presentation (15 points)

- Exemplary (13-15): The presentation is clear, engaging, and persuasively communicates the problem, solution, and potential impact. Visual aids and delivery are highly effective.
- Proficient (10-12): The presentation is mostly clear and effectively communicates the key aspects of the project. Visual aids and delivery are adequate.
- Developing (7-9): The presentation lacks clarity or fails to effectively communicate some key aspects of the project. Visual aids and delivery may need improvement.
- Beginning (0-6): The presentation is unclear, unpersuasive, or fails to communicate the project's essential elements. Visual aids and delivery are ineffective.

Total Points: 100

THE AI CREATIVITY SHOWCASE: A PROJECT-BASED LEARNING GUIDE

Objective: The AI Creativity Showcase is a project-based learning activity designed to engage students in exploring the creative possibilities of generative AI tools for artistic expression. By participating in this showcase, students will gain hands-on experience using AI models for text, image, music, and audio generation while also developing an understanding of ethical considerations and responsible AI use in creative contexts.

Materials:

- Introduction to generative AI presentation
- Access to generative AI tools for text, image, music, and audio
- Ethical guidelines for AI use in creative projects
- Project proposal template
- Creative journal template
- Showcase curation guidelines
- Presentation and Peer Critique Guidelines

Step-by-Step Guide:

Introduction to Generative AI (1-2 class sessions)

- Provide an overview of different AI models for text, image, music/audio generation, and their creative capabilities.
- Showcase examples of AI-generated art, stories, and music to inspire students.

Ethical Considerations (1 class session)

- Discuss intellectual property, bias, and responsible AI use in creative contexts.
- Establish guidelines, such as disclosing AI assistance and avoiding harmful content.

Ideation (1-2 class sessions)

- Students brainstorm creative project ideas they want to explore using AI tools (e.g., stories, artwork, songs).
- Encourage students to think outside the box and consider unique applications of AI in their creative fields.

Tool Exploration (2-3 class sessions)

- Give students time to experiment with different generative AI applications and learn prompt engineering techniques.
- Provide guidance and support as students familiarize themselves with the tools.

Project Proposals (1 class session)

- Students submit a proposal outlining their creative vision, which AI tools they plan to use, and their intended process.
- Provide feedback and suggestions to help students refine their project ideas.

Creation Phase (4-6 class sessions)

- With guidance, students iteratively generate and refine AI outputs through prompt refinement to produce their final pieces.
- Encourage students to experiment, take risks, and push the boundaries of what's possible with AI-assisted creativity.

Documentation (ongoing throughout the project)

- Students maintain a creative journal chronicling their process, challenges, and reflections on using AI.
- Encourage students to document their prompt iterations, AI outputs, and creative decisions.

Showcase Curation (1-2 class sessions)

- Select exemplary student projects to feature in a gallery exhibition, reading, or performance.
- Work with students to curate a diverse and engaging showcase highlighting the range of AI-assisted creative work.

Presentations (1-2 class sessions)

- Students present their final works to peers, explaining their creative process and AI tool usage.

- Encourage students to share their challenges, successes, and lessons learned throughout the project.

Peer Critique (1 class session)

- Provide constructive feedback on creativity, technical skills with prompts, and responsible AI integration.
- Foster a supportive and inclusive environment where students learn from each other's experiences and perspectives.

AI Creativity Showcase Marking Rubric

Creativity and Originality of the AI-Assisted Work (25 points)

- Exemplary (21-25): The AI-assisted artwork, story or music demonstrates exceptional creativity, originality, and innovative use of AI tools.
- Proficient (16-20): The AI-assisted work shows good creativity and originality, with some innovative elements in using AI tools.
- Developing (11-15): The AI-assisted work displays some creative elements but may lack originality or innovation in using AI tools.
- Beginning (0-10): The AI-assisted work lacks creativity, originality, or innovative use of AI tools.

Technical Proficiency in Using Generative AI Tools and Prompt Engineering (20 points)

- Exemplary (17-20): The student demonstrates a high level of proficiency in using generative AI tools and prompt engineering techniques to achieve their creative vision.
- Proficient (13-16): The student understands generative AI tools and prompt engineering, with room for improvement in achieving their creative goals.
- Developing (9-12): The student displays a basic understanding of generative AI tools and prompt engineering but may struggle to use them effectively to realize their creative vision.
- Beginning (0-8): The student demonstrates limited proficiency in using generative AI tools and prompt engineering, resulting in work that fails to meet their creative goals.

Responsibility and Ethics in AI Use (20 points)

- Exemplary (17-20): The student consistently demonstrates responsible and ethical use of AI, including proper attribution, disclosure of AI assistance, and avoiding harmful content.
- Proficient (13-16): The student generally adheres to responsible and ethical AI practices but may have minor lapses in attribution, disclosure, or content considerations.
- Developing (9-12): The student shows some understanding of responsible and ethical AI use but may have significant gaps in attribution, disclosure, or content considerations.
- Beginning (0-8): The student demonstrates a lack of understanding or adherence to responsible and ethical AI practices, with serious issues in attribution, disclosure, or content considerations.

Depth of Reflection and Documentation in the Creative Journal (20 points)

- Exemplary (17-20): The creative journal provides an insightful reflection on the student's creative process, challenges, successes, and lessons learned using AI tools.
- Proficient (13-16): The creative journal offers a solid reflection on the student's creative journey, with some insights into their process, challenges, and learnings, but may lack depth in certain areas.
- Developing (9-12): The creative journal provides a basic account of the student's creative process but may lack sufficient reflection, insights, or documentation of their use of AI tools.
- Beginning (0-8): The creative journal is incomplete, superficial, or fails to adequately document and reflect on the student's creative process and use of AI tools.

Clarity and Effectiveness of the Final Presentation and Peer Critique Participation (15 points)

- Exemplary (13-15): The student delivers a clear, engaging, and informative presentation that effectively communicates their creative process, AI tool usage, and final work. The student actively participates in peer critiques, offering constructive feedback and insights.

- Proficient (10-12): The student delivers a clear, informative presentation that adequately conveys their creative journey and use of AI tools. The student participates in peer critiques but may not always provide in-depth feedback.
- Developing (7-9): The student's presentation lacks clarity or fails to communicate some aspects of their creative process and AI tool usage. The student's participation in peer critiques is limited or superficial.
- Beginning (0-6): The student's presentation is unclear, uninformative, or fails to convey their creative journey and use of AI tools adequately. The student does not actively participate in peer critiques or provide unhelpful feedback.

Total Points: 100

4

GENERATIVE AI TRANSFORMING EDUCATION WITH CREATIVE AND INTERACTIVE TOOLS

I magine you're in a classroom where your students engage in lively debates with an AI, sharpening their critical thinking skills and exploring complex topics in unprecedented depth. Picture yourself planning immersive field trips and crafting personalized learning experiences with the help of an assistant, all while saving hours of your time. You can harness this transformative potential of generative AI in education.

As an educator, parent, or school administrator, you know the transformative potential of generative AI for the education sector. Systems like ChatGPT are revolutionizing how you facilitate interactive, student-centered learning by generating new content such as text, images, music, and more. With these tools, you are not just imparting knowledge but also unlocking unprecedented creative opportunities for students.

In this chapter, you'll discover how real-world examples bring the potential of AI to life. First, you'll see how a high school teacher uses ChatGPT to take classroom debates to the next level by fostering critical thinking and ethical reasoning skills that students will carry far beyond the walls. Next, you'll explore how AI can streamline administrative tasks like field trip planning, freeing teachers to focus on what matters most: igniting a love of learning in their students.

As we journey through this chapter, you'll discover essential insights and actionable strategies to bring the transformative power of generative AI into your educational practice. Prepare to expand your horizons in teaching and learning. The revolution in education begins with you.

SIDEBAR: What is Generative AI?

Generative AI refers to artificial intelligence systems that create new, original content based on patterns learned from vast datasets, such as text, images, music, and more. This sets it apart from other types of AI that are focused on tasks like classification, prediction, or optimization.

Some key examples of generative AI include:

- Language models like ChatGPT can write essays, stories, poems, and dialogues.
- Image generators like DALL-E can create realistic images and artwork from textual descriptions.
- Music generators like Jukebox can compose new songs in various genres and styles.

Generative AI is unique because it can produce novel and coherent content, often in ways that mimic human creativity. For instance, ChatGPT can engage in freeform conversation and answer follow-up questions, while DALL-E can combine disparate concepts (like "an astronaut riding a horse") into plausible images.

However, it's important to note that generative AI does not create content based on real understanding but rather on statistical patterns in the data it was trained on. It also has limitations, it can sometimes produce biased, factually incorrect, or nonsensical content.

Generative AI has immense potential to serve as an interactive learning tool and creative partner for students and teachers in education. But it must be used thoughtfully and in combination with human judgment, it is not a replacement for the knowledge and skills educators bring.

CASE STUDY 1: ENHANCING CRITICAL THINKING THROUGH DEBATE WITH CHATGPT

Imagine you're in Mr. Johnson's shoes, a high school social studies teacher passionate about igniting lively debates among your students. Climate change is on the agenda today, a topic guaranteed to spark curiosity and fierce discussion. But this time, you have a new tool up your sleeve to elevate the experience: ChatGPT.

As you introduce ChatGPT to your class, you can see the excitement and intrigue in their eyes. You explain how this AI marvel can help them dive deeper into the complexities of climate change, unearthing a treasure trove of arguments, evidence, and counterarguments. It's like having an instant research assistant at their fingertips.

But you know this is about more than just collecting information. It's a chance for your students to sharpen their critical thinking skills in a world increasingly shaped by AI. As they engage with ChatGPT, you encourage them to think analytically. You want them to consider questions like, "Is this fact or opinion? What's the source? Is there any bias lurking beneath the surface?" You want them to consider these questions as they navigate the AI-generated content.

And they navigate. You watch with pride as your students plunge into the debate, armed with the knowledge and insights gleaned from their ChatGPT-assisted research. They construct compelling arguments, anticipate counterpoints, and engage in lively, respectful discussions. It's impressive to see.

During the debrief, you witness the real transformation. You guide your students through a reflection on their experience with ChatGPT, and you see the gears turning in their minds. They share how it expanded their perspectives, exposed them to new ideas, and challenged their assumptions. They reflect on the importance of critically evaluating information, regardless of source. And they grapple with the ethical implications of AI in our lives.

You realize this is the true power of integrating ChatGPT into your classroom debates. It's not just about the content knowledge gained, impressive as that may be. It's about the life-long skills your students are developing, the ability to think critically, to engage with technology responsibly, and to navigate the complexities of an AI-driven world with discernment and integrity.

Wrapping up the lesson, there's an undeniable sense of excitement for what lies ahead. With innovative tools like ChatGPT at your disposal, you're empowered to transform classroom debates into dynamic, immersive learning experiences. You have the opportunity to equip your students with the skills they need to thrive in a rapidly changing world and foster a love of learning that will last a lifetime.

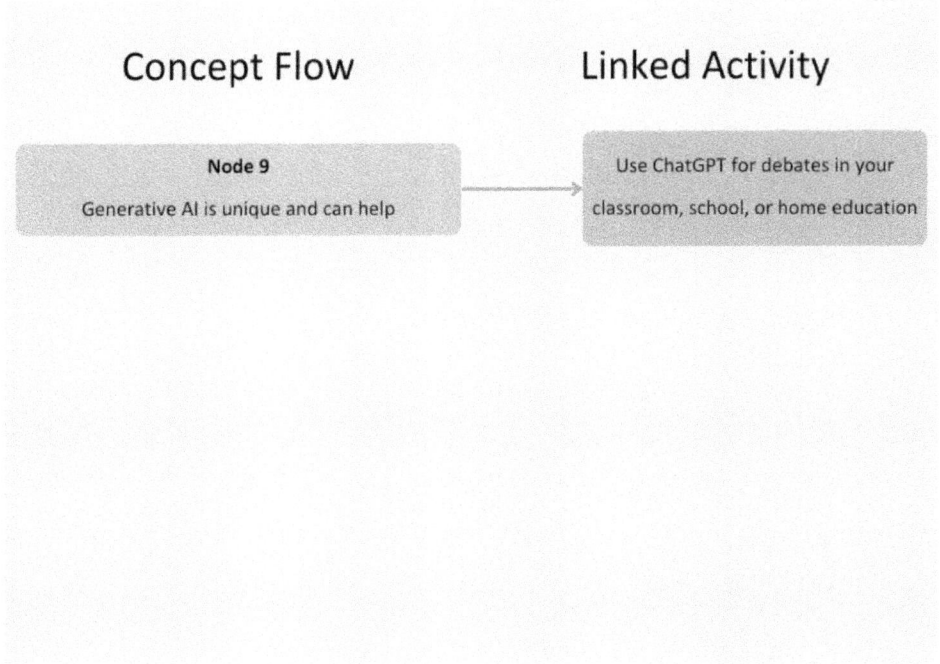

Step-by-step guide to implementing ChatGPT for debates in your classroom, school, or home education:

1. Introduce ChatGPT to your students and explain its functionality. Show them examples of it in action, highlighting its capabilities and limitations. This will ensure that they have a clear understanding of what they can expect.

2. Define the debate format you want to use. Decide whether students will debate against each other or interact with ChatGPT.

3. Establishing clear rules and structure for the debates is important to ensure a productive and organized discussion.

4. Select topics that are relevant and thought-provoking for your students. Consider the age group and educational level of your students when choosing these. This will help ensure that the debates are engaging and meaningful.

5. Climate change is an excellent choice as it is a complex, multifaceted issue that provides rich discussion.

6. Encourage students to conduct thorough research on their chosen topics. They can use ChatGPT to gather initial information and ask questions to deepen their understanding. This will help them develop well-informed arguments.

7. Remind them to consult other reputable sources to verify information and get a balanced perspective.

8. Provide opportunities for students to practice using ChatGPT. They can use it to test their arguments and receive feedback on the coherence and persuasiveness of their points. This will help them refine their skills before the actual debates.

9. When organizing the debates, make sure to monitor the interactions if you're using ChatGPT as an opponent. This will help ensure that the debates stay productive and educational.

10. Encourage students to participate and engage in constructive discussions actively.

11. After the debates, take time to reflect and review with your students. Discuss what they have learned and encourage them to evaluate the strengths and weaknesses of the arguments presented. Also, discuss the role that ChatGPT contributed to the overall learning experience.

12. Use the insights gained from these sessions to improve your approach continuously. Consider better integrating ChatGPT into future debates or other learning activities. This will help you make the most of this tool and create a dynamic and interactive learning environment for your students.

ETHICAL CONSIDERATIONS AND BEST PRACTICES

In Mr. Johnson's class, you saw students delve into the debate topic, wrestling with the moral implications of AI tool usage. This is vital in our digital age. They honed their skills in critiquing AI-generated content, leveraging it to strengthen their research, and, above all, using technology judiciously and ethically.

Integrating AI tools such as ChatGPT into education necessitates your guidance. You must help students recognize biases and verify facts to ensure responsible and ethical use. By embracing these technologies, you must foster an understanding of ethical AI, an essential part of their learning journey.

Guidelines for educators to ensure ethical use

You are responsible for shaping how AI tools are integrated into your classrooms ethically and responsibly. Here are some guidelines to help you navigate this process:

1. Encourage critical evaluation by guiding your students to ask probing questions that assess AI-generated content's accuracy, relevance, and potential biases. You might ask students: "What sources did the AI tool draw from to generate this

content? How can we verify the information it provides? What perspectives or experiences might be missing or underrepresented?"

2. Establish explicit guidelines for when and how to use AI tools in your work and understand the consequences of misuse. For example, it's important always to disclose when an AI tool has been used and to cite it appropriately, just as with any other source.

3. As an educator, you know that your own practices can set a powerful example for students. It is important to be transparent about when and how you use AI tools in your teaching. By sharing your thought process for using these tools ethically, you invite students to reflect on and discuss the implications of AI use in your subject area.

4. Keep yourself informed and engaged: The landscape of AI in education is evolving at a brisk pace, and staying abreast of the latest developments and best practices is crucial. Look for professional learning opportunities, become a part of educator communities that focus on AI integration, and engage in ongoing conversations with your colleagues to exchange insights and strategies.

By proactively implementing these guidelines, you can harness the power of AI tools to enhance student learning. You'll also cultivate the critical thinking skills and ethical frameworks students need to navigate an increasingly AI-driven world. It's an exciting and important responsibility you're well-equipped to take on together.

ADMINISTRATIVE AI CASE STUDY: USING AI TOOLS TO REDUCE ADMINISTRATIVE BURDEN

Let's say you are a teacher tasked with planning a school trip. Your mission is to craft a thorough plan, select activities stimulating learning, and ensure your student's safety. Daunting, isn't it? This undertaking often consumes time and energy, potentially diminishing the educational value.

But what if AI could intervene and provide a solution?

AI tools emerge as heroes in school trip planning. They automate scheduling, curate educational activities, and offer insider tips, transforming a difficult process into streamlined success.

Imagine you're Mrs. Roberts, a savvy teacher. You harness the power of ChatGPT to orchestrate the perfect bakery visit. Here's how AI comes to her rescue:

Information Filtering: Mrs. Roberts starts by asking ChatGPT to serve up a variety of local bakeries that offer educational experiences. In response, ChatGPT presents a curated selection. The list is rich with details, showcasing tours, workshops, and the amenities each provides.

- **Constraint Analysis:** Next, Mrs. Roberts uses ChatGPT to analyze logistical constraints like distance, group size, and accessibility. This helps her quickly narrow the contenders and find the perfect bakery match.
- **Schedule Optimization:** With the venue secured, Mrs. Roberts turns to ChatGPT to develop a suggested itinerary. The AI assistant considers travel time, activity duration, and breaks to create a balanced schedule that maximizes learning opportunities.
- **Decision Mapping:** Mrs. Roberts relies on ChatGPT to map out key decisions and contingencies as the trip takes shape. The AI tool helps her easily navigate every aspect, from permission slip management to dietary restrictions.
- **Iterative Improvement:** Finally, Mrs. Roberts uses ChatGPT to gather feedback from students and parents, using their insights to fine-tune the trip plan. The result? An educational experience that improves significantly, like a flawlessly baked croissant.

Mrs. Roberts transforms trip planning from a tedious task into an efficient, enjoyable process by wielding AI tools like ChatGPT. And the benefits don't stop there, they can also streamline tasks like:

- Crafting personalized learning plans
- Analyzing student data to identify knowledge gaps
- Generating progress reports and parent communications
- Recommending instructional strategies based on learning styles

Thanks to the power of AI, the future of educational administration is bright. As more teachers like Mrs. Roberts embrace these tools, they'll unlock precious time and energy to focus on what matters most: igniting a love of learning in their students.

GETTING STARTED WITH GENERATIVE AI IN YOUR CLASSROOM

Here is a step-by-step guide with bite-sized activities that offer low-stakes opportunities for integrating generative AI tools into your lessons. As your confidence grows with these simple exercises, feel free to expand to more complex projects and assignments!

1. Start by familiarizing yourself with the AI tool you want to use, such as ChatGPT. Explore its features, capabilities, and limitations.
2. Identify a specific learning objective or activity where the AI tool could be beneficial, like the debate activity in Mr. Johnson's class.
3. Introduce the AI tool to your students. Explain what it is, how it works, and why you use it in the classroom. Demonstrate its use with a simple example.
4. Provide clear guidelines and expectations for how students should use the AI tool. Emphasize the importance of critical thinking, fact-checking, and ethical use.
5. Integrate the AI tool into a specific lesson or activity. For example, students can use ChatGPT to research and build arguments for a debate, as Mr. Johnson did.
6. Monitor student use of the AI tool. Provide guidance and feedback as needed. Encourage students to share their experiences and any challenges they encounter.
7. Facilitate a reflection session after the activity. Have students discuss what they learned, how the AI tool helped (or hindered) their learning, and any ethical considerations that came up.
8. Gather feedback from students on their experience with the AI tool. Use this to refine your approach and plan future activities.

Exploring ways to integrate technology in the classroom can be an exciting journey. Why not try simple activities with generative AI tools like ChatGPT to enhance your comfort level and proficiency? These offer exciting opportunities for learning and engagement from which you and your students can greatly benefit.

When you're looking for new ways to inspire your students' creativity, consider using ChatGPT to generate **unique writing prompts**. You might ask for prompts like "a mystery story set in a haunted castle" or "a science fiction tale involving time travel." This tool can help you provide various options, allowing students to select a prompt that sparks their imagination for a short story, poem, or essay. It's a great way to excite your class about writing and exploring new ideas.

Encourage your students to use ChatGPT **for analysis** when you provide them with a piece of text, like a news article, short story, or historical document. They can ask questions about the main ideas, themes, the author's purpose, tone, or literary devices. For example, they might question, "What is the main theme of this short story?" or inquire, "How does the author use metaphors to convey emotion in this poem?" This activity is a great way for you to help them hone their close reading and critical thinking skills.

Vocabulary practice can become an engaging experience when you involve ChatGPT in the process. Encourage your students to input a new word and ask it to craft a sentence that illustrates the word's meaning in context. For instance, they might type, "Can you give me a sentence using the word 'diligent'?" After receiving their sentences, invite them to discuss or write about the usage. This approach reinforces acquisition and deepens their understanding of how words function within sentences.

Imagine you are the **historical figure** that your students are studying. Now, you can bring history to life using ChatGPT to engage in a role-play conversation with that figure. You can ask questions about their life, accomplishments, challenges, and perspectives. For instance, if you're delving into the Civil Rights Movement, you could have a thought-provoking conversation with Martin Luther King Jr. This activity makes history more tangible and invites you to explore the experiences and viewpoints of those who shaped our past.

Checklist for implementing generative AI in the classroom:

- Choose a specific AI tool (e.g., ChatGPT)
- Familiarize yourself with the tool's features and capabilities
- Identify a learning objective or activity where the tool could be useful
- Introduce the tool to students and provide clear guidelines for use
- Integrate the tool into a specific lesson or activity
- Monitor student use and provide guidance as needed
- Facilitate a reflection session and gather student feedback
- Adjust your approach based on feedback and plan future activities

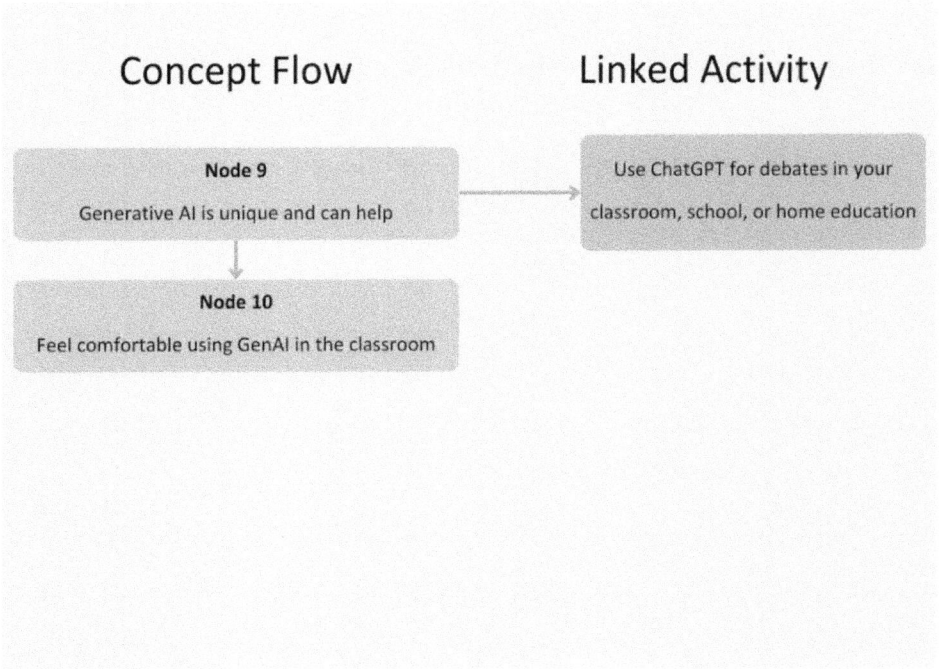

CONCLUSION: THE FUTURE OF GENERATIVE AI IN EDUCATION

Throughout this chapter, you've discovered the transformative potential of generative AI tools like ChatGPT in education. Whether you're enhancing classroom debates, fostering critical thinking skills, streamlining administrative tasks, or enabling more immersive learning experiences, they are reshaping what's possible in your teaching and learning environment.

Generative AI holds untapped potential that you will find fascinating. As this technology evolves, you can anticipate the arrival of even more sophisticated and user-friendly tools. Picture AI systems tailored to design unique learning journeys for every student, adjusting the material and progression to their individual styles and interests as they go. Envision virtual learning spaces where students can engage with historical icons, traverse the expanse of distant planets, and work together on projects with classmates worldwide, with generative AI at the helm.

The future of education combines AI and human creativity to help every student reach their potential. As a teacher, you're crucial to making this happen by

- using these tools in your classes,
- sharing what works and what doesn't with other teachers,
- and always putting your students first.

As a teacher, you can leverage generative AI to make learning experiences more captivating, meaningful, and fair for everyone. By working together, you can help build a future where every student has the opportunity to thrive, aided by a touch of a little (artificial) intelligence.

Cheat Sheet

Debate with ChatGPT: Unleashing Critical Thinking and Ethical AI Use in the Classroom

1. Choose a debate topic.
2. Introduce ChatGPT to students.
3. Demonstrate ChatGPT's use.
4. Split students into groups.
5. Conduct research with ChatGPT.
6. Develop arguments and rebuttals.
7. Organize the debate.
8. Facilitate a reflection session.
9. Provide feedback.
10. Discuss ethical considerations.

Streamlining Field Trip Planning: Harnessing AI Tools to Simplify and Enhance the Experience

1. Information filtering.
2. Visual representations.
3. Constraint analysis.
4. Schedule optimization.
5. Decision mapping.
6. Iterative improvement.

THE AI COMFORT ZONE

UNDERSTANDING AND OVERCOMING RESISTANCE TO CHANGE

PSYCHOLOGICAL BARRIERS TO TECHNOLOGY ADOPTION

Embracing change can be an emotional rollercoaster, can't it? It's normal to feel hesitant or even fearful when faced with unfamiliar territory, like integrating AI into our classrooms. After all, we're creatures of habit who find comfort in our well-worn routines.

While change can be daunting, it also opens up a world of exciting possibilities. By understanding the psychological barriers we face, as well as both the emotional and practical challenges, we can develop strategies to overcome them.

Think about a time when you took a leap of faith and tried something new in your teaching. Maybe it was a fresh lesson plan or an innovative technology. It probably felt a bit overwhelming at first, right? But as you gained familiarity and saw the positive impact on your students, that discomfort likely transformed into confidence and enthusiasm.

The journey of integrating artificial intelligence into education is similar. It may feel like uncharted territory now, but by supporting each other through the challenges, we can pave the way for a future where AI enhances and empowers our teaching.

As we explore the common psychological barriers to AI adoption, from fear of the unknown to concerns about job security, remember that you're not alone. Together, we can navigate this new landscape and create an AI-powered classroom that brings out the best in our students and ourselves. Are you ready to take that first step?

DEALING WITH A SENSE OF OVERWHELM

Feeling overwhelmed by rapid changes is a common experience known as change fatigue. This phenomenon is marked by feelings of stress, exhaustion, and powerlessness that arise from continuous shifts in educational practices and expectations. The sense of being overwhelmed often stems from uncertainty and ambivalence about these changes, especially when they occur rapidly and without clear communication or adequate support.

Rapid changes in schools can heighten stress levels among teachers, administrators, and even parents, leading to anxiety and burnout. This was particularly evident during the shifts required by the COVID-19 pandemic, which introduced new challenges and intensified existing stresses in educational settings.

A lack of preparedness for these changes and insufficient communication about the reasons and benefits behind new technologies like AI can exacerbate these feelings of overwhelm.

Strategies to Manage Overwhelm:

- Seek Information: Actively seek clear, concise information about the implemented changes. Understanding the rationale and potential benefits can help reduce uncertainty and anxiety.
- Prioritize Self-Care: Engage in activities that promote relaxation and stress reduction, such as exercise, mindfulness practices, or hobbies you enjoy. Taking care of your well-being is crucial during times of change.
- Connect with Others: Share your experiences and concerns with colleagues who may be facing similar challenges. Developing a support network can provide a sense of camaraderie and validation.
- Focus on What You Can Control: While you may not be able to control the changes themselves, focus on what you can control, such as your attitude, skill development, and classroom management strategies.

By acknowledging the sense of overwhelm and implementing strategies to manage it, educators can navigate the challenges of AI integration with greater resilience and adaptability.

FEAR OF OBSOLESCENCE AND PUBLIC PERFORMANCE ANXIETY

Many educators feel uneasy. The rapid pace of technological change can lead to fears of being left behind or even replaced by intelligent machines. This anxiety is compounded by the pressure to perform in front of students while navigating unfamiliar digital tools.

We have already extensively covered the fact that AI is not designed to make teachers obsolete. AI can be a powerful ally in enhancing the educational experience and easing the administrative burden.

Of course, integrating AI into the classroom requires a willingness to learn and adapt. Educators must proactively seek professional development opportunities to stay current with emerging technologies. Ultimately, the key to overcoming fears of obsolescence and performance anxiety lies in shifting our perspective. Educators can harness AI's potential to create more engaging, effective learning experiences by viewing it as a collaborator rather than a competitor. With an open mind and a commitment to ongoing growth, teachers can confidently navigate the ever-evolving landscape of educational technology.

DEPERSONALIZATION OF EDUCATION

Many educators worry that an increased reliance on technology might diminish the vital personal connection between teachers and students (Wong, 2023). This concern is deeply rooted in the fear that technology could transform the classroom environment into a less human-centric space.

Research indicates that depersonalization, which can lead to a sense of detachment from one's work and the people one serves, is closely linked to burnout among educators. Burnout is a significant issue characterized by emotional exhaustion, depersonalization, and a diminished sense of personal accomplishment. The ongoing challenges posed by the COVID-19 pandemic, such as adapting to new teaching methods and tools, have only intensified these feelings of burnout, making the fear of depersonalization even more pressing.

However, AI can be leveraged to strengthen, rather than weaken, personal connections in educational settings. For instance, AI-powered adaptive learning systems can provide personalized feedback and support to each student, allowing teachers to focus on building meaningful relationships. Additionally, AI can facilitate more efficient communication between teachers, students, and parents, enabling more frequent and targeted interactions.

Educators can enhance the human elements of teaching and learning by thoughtfully integrating AI into educational practices rather than diminishing them. This requires a proactive approach that prioritizes developing and maintaining strong interpersonal connections, even as new technologies are adopted.

'CONNECT TO YOUR WHY'

Feeling disconnected from your purpose as an educator while navigating the challenges of AI integration? Reconnecting with your 'why' can be a powerful way to stay motivated and focused. This quick exercise will help you align your daily efforts with your core values and aspirations.

Here's how to 'Connect to Your Why':

- Reflect on your core values: What principles guide your life and work?
- Remember your initial inspiration: What drove you to become an educator?
- Link daily tasks to your values: How does each activity connect to your 'why'?
- Identify areas for realignment: Which tasks don't align with your purpose?
- Create a personal mission statement: Summarize how your work fulfills your 'why'.
- Keep your 'why' visible: Post reminders to stay connected to your purpose.
- Regularly reassess: Ensure your actions continue to align with your 'why'.

By grounding yourself in your 'why', you'll find the resilience and clarity to embrace the opportunities AI presents while staying true to your core purpose as an educator. So, take a moment to reconnect with your 'why', your students and your own fulfillment depend on it!

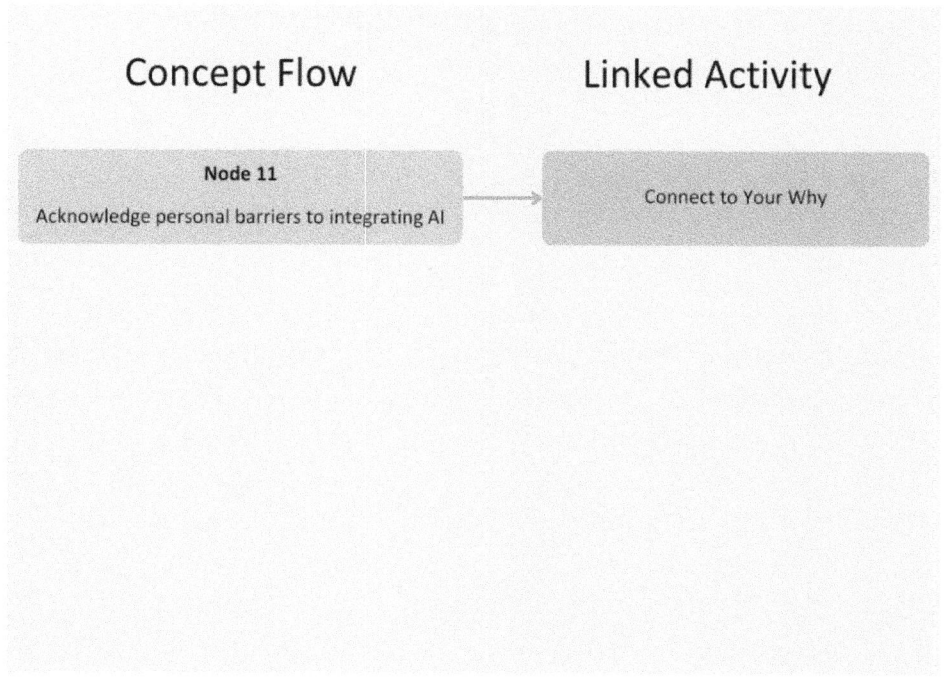

PERCEIVED LACK OF EFFICACY

Many educators are understandably skeptical about how much technology can enhance teaching and enrich student learning. The heart of their concern? The perceived gap between hands-on, face-to-face teaching and the sometimes impersonal nature of technology. This is particularly poignant in settings where resources are scarce, and teachers feel underequipped due to a lack of proper tools and training.

Despite these hurdles, the drive to integrate technology in classrooms continues, fueled by a strong belief in its potential to revolutionize learning experiences. To tap into this potential, a balanced approach is crucial. This includes continuous professional development to empower teachers and ensure that every student has equal access to these technological tools. Together, these steps can help bridge the gap between technology and effective teaching, making the digital leap less daunting and more rewarding.

Opportunities for Empowerment:

Imagine the possibilities when AI is harnessed to supercharge teaching effectiveness. Adaptive learning platforms like Knewton can provide personalized content for each student, freeing teachers to offer more targeted support. As one educator shared, "AI has given me precious time to focus on what matters most - connecting with my students and helping them grow."

Moreover, AI-powered tools like Gradescope can automate grading tasks, allowing teachers to provide more detailed, individualized feedback. "It's been a game-changer," remarked another teacher. "I can now give each student the attention they deserve without sacrificing my own well-being."

By embracing AI as an empowerment tool, educators can transform perceived limitations into opportunities for unparalleled effectiveness and deeper student engagement.

RELUCTANCE TO ALLOCATE TIME

Teachers in K-12 education often exhibit reluctance to allocate time towards learning about new technologies, a hesitation influenced by the perceived increase in workload. This reluctance can stem from various factors, including the complexity of technology, fear of failure, lack of confidence, and general anxiety about using new tools. Some educators may cling to traditional teaching methods due to these concerns, leading to the underutilization of available educational technologies.

The perceived burden of incorporating new technologies into existing teaching practices can be daunting, making teachers feel overwhelmed. Factors such as a lack of knowledge, skills, and confidence in using technology and concerns about privacy further fuel this reluctance. Teachers who lack confidence in their ability to use educational technologies may experience higher stress levels, underscoring the need for adequate support and training.

Moreover, the perceived benefits versus the costs of adopting new technology significantly influence teachers' decision-making processes. Concerns about the financial implications and the pressure to keep up with technological advancements can contribute to their hesitation.

Strategies for Time-Efficient AI Integration:

- Start small: Dedicate 10-15 minutes daily to exploring AI tools specifically designed for educators. Consistency is key to building familiarity and confidence.
- Prioritize relevance: Focus on AI applications that directly address your most pressing needs, such as grading assistance or personalized learning recommendations.

- Collaborate with peers: Share resources and experiences with colleagues who are also interested in AI. Learning together can make the process more efficient and enjoyable.

Educators can gradually overcome their reluctance and unlock the long-term benefits of these powerful tools by taking small, purposeful steps toward AI integration.

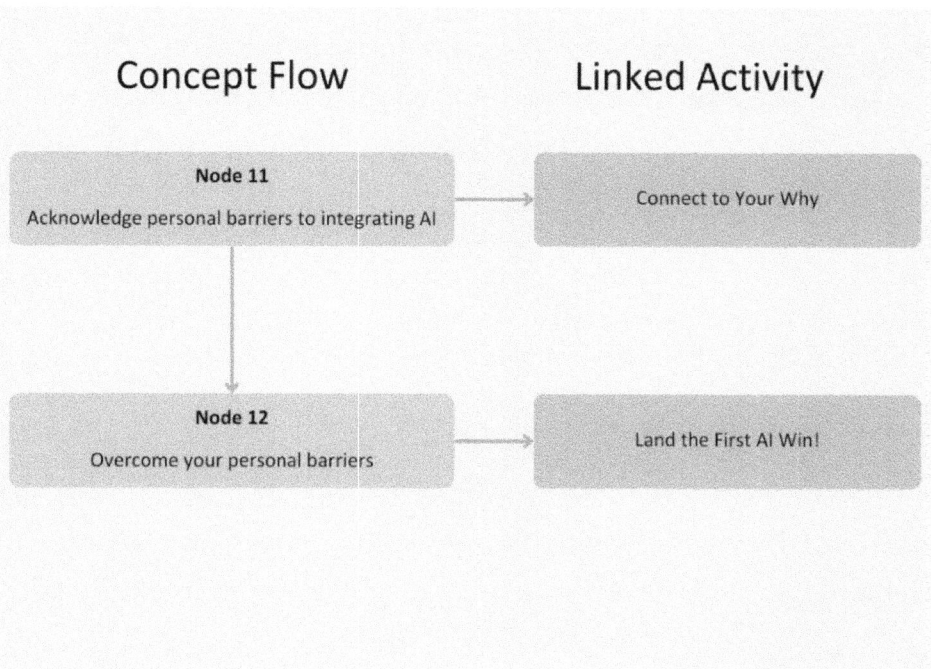

LANDING THE FIRST AI WIN!

Looking for a quick win with AI? Start by automating a simple task, like attendance tracking. Here's how: Choose an AI tool recommended by your peers, test it in one class, and observe the results. Celebrating this small success can boost your confidence to take on more ambitious AI projects.

These early victories boost **the morale of educators and students alike.** They also serve as tangible proof of the potential benefits that technology can bring to the learning environment.

For instance, a study on Information Technology (IT) adoption in small businesses revealed that successful initial implementations could enhance educators' confidence and willingness to explore and utilize technology in their teaching practices.

Furthermore, the psychological impact of small wins cannot be overstated. They act as milestones that validate the effort invested in adopting new technologies, thus encouraging continued engagement and experimentation.

In education, where resistance to change and technology anxiety are common, acknowledging and celebrating these small wins can lead to a cultural shift towards a more innovative and technologically receptive environment. This is supported by findings that suggest that needs-based technology integration education, tailored to the specific requirements of educators, can lead to rapid positive changes in attitudes towards technology. Ultimately, this benefits student learning outcomes. Therefore, recognizing and building upon small wins is not just a strategy but a necessary step in the successful and sustainable integration of AI and other emerging technologies into K-12 education. It inspires educators to navigate the complexities of digital transformation with confidence and purpose.

HOW TO GET THAT FIRST WIN!

Step 1: Identify a Minor Pain Point

A small win calls for a small pain point to be overcome! Fortunately, education is full of potential small wins. So, you should look for repetitive low-value tasks like taking attendance, grading multiple-choice questions, and organizing files that take up your time.

Step 2: Research AI Solutions

To identify the right AI solutions for your classroom, explore the pre-approved tools available in your school's technology marketplace. This allows you to integrate solutions that are compatible with existing infrastructure and policies easily. Additionally, you can expand your search by attending webinars or conferences on AI in education. There, you can discover new capabilities and get hands-on experience. Finally, don't underestimate the value of reaching out to peers already using AI in their teaching. They can provide invaluable advice on real-world implementation, pitfalls to avoid, and tips for getting started. Moreover, a simple conversation over coffee with an AI trailblazer from another school could spark creative ideas for landing your first classroom win!

3. Start Small, Start Soon

Identify a minor pain point in your classroom to achieve that first AI win. For example, you can choose something like taking attendance or grading multiple-choice questions, which can take up precious time. Then, research easy-to-use AI solutions that require little financial investment to address that need. Once you have found a suitable tool, ensure it seamlessly integrates with your existing systems. Next, pilot the tool for a short period with just one class. During this pilot phase, focus more on learning and getting comfortable with the technology rather than demanding perfection immediately. Once you have tested it and gained experience, you can expand the tool's usage to more classes.

Additionally, gather metrics on time and efficiency gains and collect student feedback to measure the impact of the AI solution. The goal is to start small but make steady progress toward integrating AI in a practical way that enhances your teaching. Throughout the process, persistence is vital in overcoming any hurdles that may arise. Remember to maintain focus on your long-term pedagogical goals. By achieving these small wins and staying committed, you can eventually achieve widespread adoption of AI in your classroom.

4. Gather Metrics

When integrating AI in education, knowing when you're winning is essential. And the only way to do that is by measuring your success! So, what should you measure?

- **Track time savings and efficiency boosts**: Note precise time gains in tasks like taking attendance or grading tests. Then, calculate the percentage differences in time required before and after AI integration.
- **Survey students:** Distribute anonymous polls and questionnaires to students. Gather testimonials and feedback on their AI learning experience. Look for themes around engagement, interest, and perceived effectiveness.
- **Analyze performance data:** Pull statistics on metrics like assignment scores, testing accuracy, and completion rates. Compare metrics between classes using AI tools and those relying solely on traditional methods.
- **Share results:** Compile metrics into easy-to-understand visualizations and highlight student quotes and experiences. Then, showcase findings at faculty meetings or in teacher communities. Quantifiable wins speak volumes, sparking further AI adoption.

Remember, measuring success is not just about the numbers. It's about understanding the real-world impact of AI integration on your teaching practices and student learning outcomes. The key is to track tangible metrics across **time, efficiency, student perceptions, and performance**. Measurements lend credibility and allow you to identify classroom victories, whether small or large, conclusively.

5. Share Your Story

Share your AI success story far and wide to motivate and inspire your colleagues. Also, showcase your metrics and student testimonials at faculty meetings. Furthermore, post highlights on teacher message boards and social media. Emphasize the tangible wins and student perspectives to spark interest in AI adoption among your peers. Additionally, lead by example, let your small victory fuel broader change.

6. Build Grit

Educators must embrace grit to persist through inevitable hurdles when they embark on an AI integration journey. Moreover, they should seek a supportive peer community to troubleshoot challenges and draw motivation from each other's small wins. Additionally, they should focus on the long-term pedagogical goals, enhanced efficiency and student learning. They should let the momentum of those first classroom victories fuel their motivation to continue experimenting with and refining the usage of AI tools. Consequently, with grit and collaboration, each minor success paves the path for the next, culminating in widespread adoption and tangible educational impacts. Finally, they should approach AI integration as a marathon, not a sprint.

EMERGING POSSIBILITIES: FROM LITTLE WINS TO COLLABORATIVE GROWTH

After achieving initial successes with AI, it's time to expand its use. This isn't a solo mission; it requires collaboration and knowledge-sharing among educators.

Building a Community of Innovators

Think of the process as a community garden. Each success with AI is a seed that grows into a thriving innovation ecosystem when shared. Here's how to cultivate this community:

- **Share Your Successes**: If you've used AI to tailor reading assignments to each student's level and it's sparked enthusiasm for reading, share this experience. It could inspire others.
- **Find Mentors**: Look for educators who have successfully integrated AI. Their experiences can guide those who are new to the process.

- **Hold Local Forums**: Create spaces where educators can share their experiences with AI integration. These can be places where practical strategies and innovative ideas flourish.

Pilot Projects: Testing the Waters

Pilot projects are like experimental garden plots. They allow you to test different AI tools and strategies to see what works best.

- **Recruit Volunteers**: Volunteers for pilot projects are like pioneers, ready to explore the unknown. Their feedback is invaluable for future planning.
- **Learn from Every Attempt**: Not every pilot project will succeed, but each provides valuable lessons. Sharing both successes and setbacks can help future efforts.
- **Stay Adaptable**: Educators should remain flexible in their AI strategies, ready to evolve with new insights and technologies, just as gardeners adjust their techniques with the changing seasons.

From Individual Efforts to Collective Success

As we proposed, AI integration in education resembles a shared garden. It starts with individual efforts but grows into a thriving ecosystem of shared success. By working together openly and thinking innovatively, we can ensure AI improves learning outcomes for every student.

CHEAT SHEET: YOUR ROADMAP TO AI INTEGRATION

New Key Terms: Implementation Strategy, Professional Development, Technology Roadmap

To improve is to change; to be perfect is to change often.

— WINSTON CHURCHILL

As we explored earlier, integrating AI can surface inner barriers such as feeling overwhelmed or lacking technical skills. However, progress requires moving through such obstacles.

Therefore, this practical checklist empowers educators to evaluate readiness, access resources, start small with low-risk pilots, gather student feedback, and lean on peer support. Moreover, by methodically addressing hurdles with grit and determination, each small win achieved with AI builds confidence, know-how, and momentum for broader adoption. Equipped with this roadmap for change, let's embark on the first steps toward AI integration victories!

Checklist content:

Assess Readiness

- Evaluate technical skills and comfort level with AI technologies; be honest!
- Survey teachers on perceived self-efficacy with AI tools and students' learning preferences.

Gather Resources

- Attend AI integration workshops and webinars at school or in the district.
- Set up a peer mentoring program to learn from colleagues who are early AI adopters.
- Explore AI tools that are pre-approved and available through the school marketplace or the IT department.

Start Small

- Identify a repetitive, low-value task (e.g., taking attendance).
- Pilot a simple AI solution with one class for a short period.

Measure Outcomes

- Note precise time savings and efficiency gains.
- Gather student feedback via anonymous polls or questionnaires
- Share metrics and student perspectives at faculty meetings.

Persist with Grit

- Seek out communities to collaborate with and motivate each other.
- Maintain focus on enhancing student learning.

- Let the momentum of small wins fuel your motivation.

This comprehensive checklist equips educators with a roadmap for overcoming obstacles like anxiety or skills gaps when integrating AI. Teachers can systematically address hurdles by taking the time to assess readiness and tapping into available resources. They can start small with low-risk pilots, gather student feedback, and lean on peer support communities. They can equip themselves with grit and determination to persist through challenges, build know-how and self-efficacy through small wins with AI, and pave the path for the next victory.

As trailblazer Grace Hopper remarked,

The most dangerous phrase is 'we've always done it that way.'

MAKE A DIFFERENCE WITH YOUR REVIEW!

UNLOCK THE POWER OF GENEROSITY!

"Education is the most powerful weapon which you can use to change the world."

— NELSON MANDELA

When we give without expecting anything in return, we create ripples of kindness and inspiration that reach farther than we can imagine. ***Let's join hands and make a difference together!***

*Would you help someone just like you—**curious about fearlessly using AI in the classroom** but unsure where to begin?*

*My mission with **Fearlessly Use AI in the Classroom: Easy-to-use Prompts, Demystify AI With a Stress-Free Guide to Essential Strategies and Step-by-Step Techniques for Transformational Classroom Instruction** is to make **AI in education accessible, fun, and empowering** for teachers, administrators, parents, and everyone who believes in the power of learning.*

*But to reach more people, **I need your help.***

*Most people choose books based on **reviews.** That's where you come in. By leaving a **review,** you can help a fellow **educator, parent, or curious learner** start their journey toward **embracing AI in their classrooms** with confidence and ease.*

*Your **review** could:*

- *Inspire **one more teacher** to try AI tools and unlock new possibilities for their students.*
- *Encourage **one more parent** to understand and support their child's learning journey.*
- *Motivate **one more school administrator** to create forward-thinking environments.*
- *Transform **one more classroom** into a hub of creativity, curiosity, and discovery.*

*All it takes is **a moment of your time** and a few kind words. It's **free,** it's **easy,** and it could mean the world to someone who's looking for guidance.*

*To leave a **review**, simply scan the **QR code** or click the link below:*

https://www.amazon.com/review/review-your-purchases/?asin=BOOKASIN

*If you love **helping others,** you're my kind of person. **Thank you** from the bottom of my heart for being part of this mission to make **AI in education approachable and transformative** for all.*

Regards,
Michael P. West

6

SETTING THE STAGE FOR AI INTEGRATION IN EDUCATION

INTRODUCTION TO TPACK

Technological Pedagogical Content Knowledge (TPACK) helps teachers understand how to integrate technology into their teaching. It's based on the idea that to create engaging and meaningful learning experiences, teachers must have a strong grasp of three key areas: technology, pedagogy, and content. This framework is often compared to a three-legged stool, if any leg is weak or missing, the stool topples.

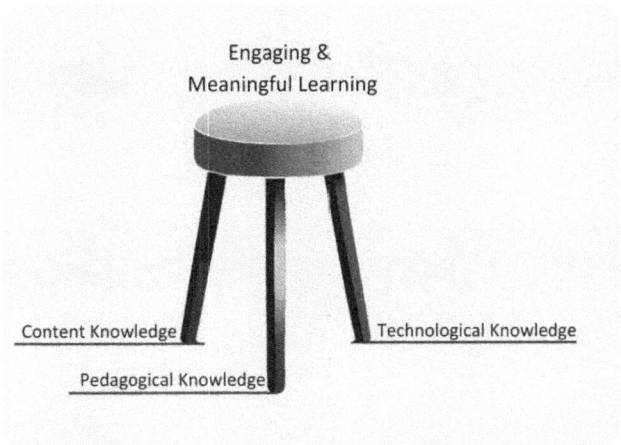

The TPACK framework emphasizes that teachers should not be experts in just one or two areas. Instead, they must understand how all three interact and influence each other. For example, a history teacher with knowledge about World War II (content knowledge) who understands how to create interactive lessons (pedagogical knowledge) might use a virtual reality app (technological knowledge) to help students experience what it was like to be in the trenches.

Just as a stool requires all three legs to be stable and functional, educators need competencies in all three domains to effectively integrate artificial intelligence (AI) into K-12 education. Educators can utilize AI's capabilities to improve learning experiences and outcomes by examining the dynamic interplay between these knowledge areas.

Successful AI integration in education goes beyond merely having "advanced" technology. It relies on educators' ability to skillfully merge their understanding of AI's potential with sound pedagogy and subject matter expertise. TPACK underscores that educators must cultivate proficiencies in technology, pedagogy, and content, and, crucially, in how these domains interact.

For instance, an AI-powered writing feedback tool is only effective when a teacher can interpret its suggestions through a pedagogical lens, considering the student's unique needs and the specific writing skills being developed. Similarly, an AI-driven science simulation must be thoughtfully woven into a well-designed curriculum, aligning with learning objectives and assessment strategies.

It is important to recognize that some educators may naturally feel more comfortable with the content and pedagogical "legs" of the TPACK stool. Their expertise in subject matter and teaching strategies may be more developed than their technological knowledge. However, to create a stable and effective learning environment that incorporates AI, educators must strive to strengthen their technological competencies. By engaging in professional development, collaborating with technology specialists, and experimenting with AI tools, educators can bolster the technological "leg" of their TPACK stool, ensuring a balanced and robust approach to AI integration.

TPACK offers a solid basis for effectively integrating artificial intelligence (AI) into K-12 education. By considering the dynamic interplay between technological knowledge, pedagogical knowledge, and content knowledge, educators can use AI's capabilities to improve learning experiences and outcomes.

Concept Flow

Node 13

Educators need to assess their readiness to implement AI technology

Linked Activity

Complete the AI Integration Readiness Checklist

DETAILED EXPLANATION OF THE TPACK FRAMEWORK

TPACK offers a comprehensive perspective on the knowledge necessary to integrate AI into educational environments effectively. It clarifies the relationship among three essential domains:

- Technological Knowledge (TK),
- Pedagogical Knowledge (PK),
- Content Knowledge (CK).

For AI integration, TK includes understanding the capabilities and limitations of AI technologies, as well as the skills to utilize them effectively. Educators proficient in TK can discern how AI tools can support or hinder the accomplishment of specific tasks.

Pedagogical knowledge (PK) refers to educators' thorough knowledge of teaching and learning processes, instructional strategies, and student needs. With strong PK, educators can use AI's capabilities, such as personalized learning and real-time feedback, to improve student engagement and outcomes.

Content knowledge involves mastery of the subject matter being taught, enabling educators to identify AI tools that align with and enrich their discipline.

The true power of TPACK emerges at the intersections of these domains. Technological Pedagogical Knowledge (TPK) allows educators to recognize how AI technologies can facilitate specific pedagogical approaches, while Technological Content Knowledge (TCK) helps them discern how AI can contribute to their subject area. TPACK represents the synthesis of these knowledge domains, enabling educators to seamlessly integrate AI tools, teaching strategies, and content in a coherent and effective manner.

AI's integration into education also raises ethical considerations that must be addressed. Concerns around transparency, fairness, accountability, and inclusiveness arise from the potential for AI systems to make biased decisions and the lack of transparency in AI decision-making. Consequently, an ethical knowledge component becomes essential for TPACK, enabling educators to assess AI-based decisions and reduce possible negative impacts. By extending TPACK to encompass ethical assessments, educators can harness AI's transformative potential while upholding the principles of equity and social responsibility.

Recent studies have demonstrated the importance of the TPACK framework in enhancing teaching and learning outcomes, particularly when integrating artificial intelligence into education:

- Training programs: Teachers need training to learn how to use TPACK effectively in their classes. These programs help them improve their skills in using technology and teaching methods that work well together.
- Online teaching: Teachers' attitudes towards TPACK and their ability to teach online classes are crucial for the success of virtual classrooms, particularly during challenging times like the COVID-19 pandemic.
- Student benefits: Research shows that TPACK helps students learn more and develop essential 21st-century skills.
- Project-based learning: Teachers who use project-based learning and TPACK tools can better integrate technology into different subjects in meaningful ways.

TPACK, combined with ethical considerations, offers a valuable framework for educators to responsibly incorporate AI in teaching and learning, assisting in realizing its possible advantages

FICTIONAL CASE STUDY: AI INTEGRATION IN ELEMENTARY EDUCATION

Note: This fictional case study extrapolates from existing case studies, as no academic case studies specifically addressing this scenario currently exist.

Maple Avenue Elementary School, a hypothetical large urban public school, has embraced TPACK to transform its approach to education. By seamlessly integrating AI tools into its curriculum, the school aims to create personalized learning experiences that cater to the diverse needs of its student body, including those with disabilities.

At the heart of this initiative is a robust collaboration between educators, AI specialists, and parents. Teachers receive comprehensive training on using AI technologies to improve their pedagogical strategies and content delivery. AI experts work closely with the faculty to develop tailored solutions, ensuring that the implemented tools align with the school's educational objectives and ethical principles.

One of the key AI tools utilized is an educational platform that adjusts the content and pace of learning to each student's needs and abilities. This platform employs computer programs that learn from data and analyze each student's performance data, learning styles, and areas of strength or difficulty. This information is then used to generate personalized lesson plans, interactive activities, and multimedia resources that cater to individual needs. For students with disabilities, the platform offers specialized accommodations, such as text-to-speech functionality, visual aids, and customized pacing.

To mitigate the risk of algorithmic bias, the school has implemented careful reviewing practices. A diverse team of educators, AI experts, and community representatives regularly review the algorithms and training data used by the AI tools, ensuring that they are free from biases and promote inclusivity. Additionally, the school has established open lines of communication with parents, fostering open dialogues about the AI integration process and addressing any concerns or feedback.

The impact of this TPACK-driven approach has been profound. Students have demonstrated improved engagement, accelerated learning progress, and heightened self-confidence. Teachers report a deeper understanding of their student's unique learning needs and an enhanced ability to provide targeted support. Moreover, the school's commitment to ethical AI use has fostered a culture of trust and collaboration within the community.

Maple Avenue Elementary School's success story serves as a powerful example of how TPACK, combined with a strong emphasis on ethical considerations, can unlock the transformative potential of AI in education, creating inclusive and equitable learning environments that empower every student to thrive.

CHALLENGES AND ETHICAL CONSIDERATIONS

The integration of AI in education presents a myriad of challenges and ethical considerations that must be carefully navigated. From the potential for algorithmic bias to concerns surrounding data privacy and transparency, these issues have far-reaching implications for the equitable and responsible use of AI in educational settings.

One of the primary challenges lies in the opaque nature of AI decision-making processes. As AI systems become more complex, it becomes increasingly difficult to understand the rationale behind their outputs or recommendations. This lack of transparency raises concerns about the fairness and accountability of AI-driven decisions, particularly regarding sensitive matters such as student assessment or resource allocation.

Furthermore, AI systems are susceptible to perpetuating societal biases present in their training data, leading to discriminatory outcomes that can exacerbate existing inequalities. For example, if an AI-based grading system is trained on a dataset that reflects historical biases against certain demographic groups, it may unfairly penalize or favor students from those groups, compromising the principles of inclusivity and equal opportunity.

The TPACK framework offers a powerful lens through which educators can navigate these challenges and uphold ethical principles in their use of AI. By interweaving technological understanding with pedagogical and content expertise, TPACK enables educators to critically evaluate AI tools through multiple lenses. Just as a chef needs to have knowledge of ingredients (content), cooking techniques (pedagogy), and kitchen equipment (technology) to create a delicious dish, educators must develop competencies in all three domains to effectively integrate AI into their teaching practices.

Through a deep understanding of AI's capabilities and limitations, educators can assess the transparency and reliability of AI-based decisions. Coupled with their pedagogical expertise, they can determine whether the AI's recommendations align with sound educational practices and cater to their students' diverse needs.

Moreover, by integrating ethical considerations into the TPACK framework, educators can develop a heightened awareness of the potential biases and implications of AI systems. This ethical knowledge enables them to scrutinize AI tools for fairness, inclusivity, and accountability, ensuring that their implementation does not perpetuate or amplify existing societal inequities.

Ultimately, the TPACK framework empowers educators to be proactive and vigilant in their use of AI, leveraging its potential while upholding the highest ethical standards and prioritizing the well-being and equitable treatment of all students.

CASE STUDY: MRS. SANCHEZ'S SIMPLE MACHINES PROJECT

Background Mrs. Sanchez, a middle school science teacher, aimed to effectively integrate technology into her lessons to enhance student engagement and learning outcomes. She attended a professional development workshop on the TPACK framework and project-based learning (PBL).

Learning about TPACK and PBL During the workshop, Mrs. Sanchez learned about the TPACK framework, which guides teachers in selecting appropriate technologies that align with their pedagogical approaches and content.

Mrs. Sanchez was interested in how the TPACK framework complemented project-based learning pedagogy. PBL allows students to gain knowledge and skills by working for an extended period to investigate and respond to an authentic, engaging, and complex question or problem. By integrating technologies through the TPACK perspective, she could further enrich these immersive learning experiences.

Designing the Project Inspired by her new knowledge, Mrs. Sanchez redesigned her unit on simple machines using a PBL approach supported by the TPACK framework:

- For the TK component, she selected simulation software that could model the behavior of simple machines such as levers, pulleys, and inclined planes.
- Her PK told her this simulation would allow students to visualize abstract concepts and test variables in a risk-free virtual environment.

To connect TK and PK, she developed the pedagogical strategy (TPK) of having students work in teams to build and analyze simple machine simulations based on the real-world scenarios they would investigate.

This pedagogical approach aligned with Mrs. Sanchez's CK goals of teaching concepts like mechanical advantage, force, and work. By interfacing the simulation software with designing and building models (TCK), students could deepen their understanding of simple machines.

Implementing the Project Throughout the project, Mrs. Sanchez:

- Tracked her students' learning through the simulation data and modeling activities, adjusting her instruction as needed.
- Facilitated productive classroom discussions around the simulations and physical models.
- Provided timely feedback and support to student teams as they worked on their projects.

Project Outcomes In the end, her students demonstrated impressive gains in their conceptual knowledge of simple machines. More importantly, they gained valuable experience in collaboration, problem-solving, and using technology as a tool for inquiry - all critical skills for STEM fields.

Mrs. Sanchez's Growth Mrs. Sanchez's TPACK has grown significantly through this PBL experience:

- She successfully integrated a technological tool in pedagogically sound ways to teach key science content while receiving feedback from her students.
- Her technological knowledge (TK) expanded as she learned to use the simulation software and other tools students used effectively.
- Her pedagogical knowledge (PK) was elevated through designing and facilitating an inquiry-based, collaborative project.
- Her content knowledge (CK) deepened as she guided students through the research, simulation, and modeling phases, identifying misconceptions and areas needing reinforcement.
- She skillfully merged her TK, PK, and CK to create an authentic, technology-enriched learning experience aligned with her pedagogy and content goals, demonstrating strong TPACK.

Mrs. Sanchez looked forward to exploring new TPACK-guided projects to continue improving her craft as an educator.

Concept Flow

Node 13

Educators need to assess their readiness to implement AI technology

Linked Activity

Complete the AI Integration Readiness Checklist

Implement a project supported by the TPACK framework

FUTURE DIRECTIONS

The TPACK framework offers a comprehensive and forward-thinking approach to integrating AI into educational settings. It enables educators to harness the transformative potential of these technologies while upholding ethical principles and prioritizing student well-being. Nevertheless, as AI continues to evolve and become integrated into educational landscapes, several crucial factors must be addressed to ensure successful and equitable implementation.

- Technological Adaptability: Schools should budget to stay current with rapidly advancing AI technologies and update their computer systems and software to maintain compatibility with the latest advancements.
- Cost and Accessibility: The use of tools that allow software to use AI functionalities can increase expenses, potentially making it harder for all schools to adopt AI if these tools are not affordable and accessible across different regions and demographics.
- Data Privacy and Ethics: Ensuring the safety of personal information and data concerning students and schools is essential. Moreover, diligent oversight and continuous improvement are necessary to mitigate the risk of perpetuating existing biases and disparities through AI systems.

- Professional Development: Educators require substantial training and dedicated professional development programs to effectively integrate AI into their teaching practices and fully utilize its potential to enhance learning experiences.
- Infrastructure Readiness: Adequate infrastructure, including high-speed internet connectivity and AI-capable devices, is essential for the widespread adoption of AI in educational settings.
- Regulatory Landscape: The evolving regulatory framework surrounding AI in education challenges institutions to balance compliance with innovation, underscoring the need for close collaboration between educators, policymakers, and AI experts.

By taking the initiative to address these factors, the TPACK framework can serve as a reliable roadmap for educators to adapt and refine their practices as AI capabilities expand and new use cases emerge in education.

Looking ahead, the TPACK framework's inherent flexibility will be invaluable as AI capabilities expand and new use cases emerge in education. Educators can use this model to explore novel applications of AI, such as intelligent tutoring systems, adaptive learning platforms, and immersive virtual environments while maintaining a steadfast commitment to ethical and responsible implementation.

Moreover, the TPACK framework's emphasis on collaboration and community engagement will become necessary as AI integration in education progresses. By fostering partnerships among educators, AI experts, policymakers, and parents, the framework facilitates open dialogues, shared decision-making, and collective efforts to ensure that AI adoption aligns with societal values and addresses the diverse needs of all learners.

As the educational landscape continues to evolve, it is crucial that educators remain at the forefront of this transformation, embracing AI as a powerful tool while vigilantly safeguarding the principles of equity, inclusivity, and ethical conduct. The TPACK framework offers a robust pathway for navigating this journey, equipping educators to shape the future of education and release the vast potential of AI to improve learning experiences for every student.

APPENDIX: AI INTEGRATION READINESS CHECKLIST FOR SCHOOLS

Before introducing the AI Integration Readiness Checklist for schools, it's important to know the importance of thorough preparation. Integrating AI into educational settings requires a tailored approach that considers the unique needs and circumstances of each institution. This guide helps teachers, administrators, and schools evaluate their current situation and identify areas needing improvement before implementing AI technologies. By assessing readiness in key domains, stakeholders can ensure a seamless transition and increase the effectiveness of AI in educational practices.

Step 1: Decide on Your AI Integration Vision

- **Task:** Outline what AI integration success entails for your school.
- **Example:** AI could personalize education to each student's needs, enabling teachers to provide focused support, which could manifest as success.

Step 2: Identify Priorities for AI Application

- **Task:** Highlight where AI could significantly impact your school and its students.
- **Example:** Increasing student engagement through interactive, AI-powered tutoring systems.

Step 3: Assess Leadership Support for AI

- **Task:** Gauge the level of AI understanding and support among educational leaders.
- **Example:** Use surveys to determine leaders' knowledge of AI benefits and potential issues.

Step 4: Evaluate Operational AI Potential

- **Task:** Identify operational aspects that can improve with AI.
- **Example:** Streamlining administrative duties with AI tools, including recording attendance and grading.

Step 5: Assess Data Management for AI

- **Task:** Review the organization of data and its usability for AI purposes.
- **Example:** Ensure student performance data is digitized and available for AI analysis.

Step 6: Examine Technology Infrastructure

- **Task:** Check if the existing technology setup can handle AI programs.
- **Example:** Assess the school's Wi-Fi capacity for increased data usage from AI tools.

Step 7: Confirm Security Protocols

- **Task:** Ensure AI systems are secured.
- **Example:** Implement robust encryption to safeguard information in AI programs.

Step 8: Address AI Integration Risks

- **Task:** Identify and plan for possible problems with AI, such as unfairness.
- **Example:** Audit AI tools for biases in understanding different languages and plan reviews to ensure fairness.

Step 9: Ensure Policy Compliance

- **Task:** Confirm that the AI tools comply with education and privacy rules.
- **Example:** Verify AI tools' compliance with laws governing student data privacy.

Step 10: Prioritize Child and Youth Protection

- **Task:** Safeguard students in the use of AI tools.
- **Example:** Use AI content filtering to block access to inappropriate materials.

Step 11: Plan Professional Development for AI

- **Task:** Organize a workshop for teachers using AI tools properly and considering important ethical issues.
- **Example:** Conduct workshops on using AI tools to enhance teaching.

Step 12: Develop an AI Integration Roadmap

- **Task:** Draft a plan that outlines the steps to start using AI, the timeline for each step, and the responsible parties.

- **Example:** Aim to trial an AI-driven personalized learning platform in a specific grade within six months.

Conclusion and Action Steps

- **Task:** Gather information to formulate a plan to fix unprepared areas.
- **Example:** If the technology setup falls short, plan to enhance it before using AI.

APPENDIX: SIMPLE MACHINE PROJECT BRIEF USED BY MRS. SANCHEZ

Driving Question: How can we apply our understanding of simple machines to improve a community-based system or process?

Overview: In this project, you will work in teams to investigate how simple machines are used in real-world applications that impact our community. You will select a particular system or process, analyze its current use of simple machines, and then propose improvements leveraging additional simple machine concepts.

To guide your investigation, you will use simulation software to model the behavior of different simple machine setups. Your simulations, combined with the physical models you construct, will allow you to calculate mechanical advantages and other metrics.

By the end of the project, your team will deliver a presentation explaining your selected community system, your findings from the simulations and models, and a proposal with visuals for how simple machines could optimize that system.

Project Milestones:

1. Team Formation & System Selection
   - Get organized into teams of 4
   - Decide on a community system or process you want to explore (e.g., parks equipment, construction vehicles, food distribution)
   - Submit the system selection with the reason for the choice
2. Research & Simulation Phase
   - Research your selected system and identify any simple machines currently used
   - Use simulation software to model lever, pulley, wedge, wheel/axle, inclined plane, and screw scenarios
   - Track simulation data (forces, work, mechanical advantage, etc.)
   - Construct physical models representing key parts of your system

3. Analysis & Proposal Development
   - Analyze data/observations to determine how introducing additional simple machines could improve efficiency, etc.
   - Brainstorm simple machine modifications and enhancements to your system
   - Create visuals/designs representing your proposed changes
   - Develop a cost-benefit assessment for implementing your proposal
4. Presentation & Report
   - Prepare a 15-minute presentation covering:
     - Overview of your selected community system
     - Simple machines currently used and their roles
     - Simulations and modeling findings
     - Your proposal for simple machine integration
     - Cost-benefit analysis and feasibility assessment
   - Create a written report compiling all of the above components

7

SIMPLIFYING ABSTRACT CONCEPTS ACROSS DISCIPLINES

THE TRANSFORMATIVE POWER OF AI IN EDUCATION

In the ever-evolving realm of education, where knowledge is cultivated and nurtured, effectively communicating abstract concepts forms the bedrock for students' comprehensive understanding, critical thinking abilities, and capacity to navigate the intricacies of the world around them. Across the diverse spectrum of subjects, from the logical domain of STEM to the creative fields of art, language, and music, it is difficult to overstate the importance of how educators are tasked with the formidable challenge of converting complex ideas into accessible, captivating learning experiences.

Nevertheless, the journey to successfully convey abstract ideas is replete with obstacles. As seen in the previous chapter, teachers must possess a robust grasp of their subject matter, coupled with the pedagogical dexterity to adapt their approaches to cater to the diverse needs of their students and bridge the chasm between abstract concepts and real-world applications. Not only that, but educators must nurture a learning environment that encourages creativity, critical reflection, and student motivation, as these elements are vital for effective engagement with abstract ideas. It's a big ask.

We want and need to show how AI can make abstract concepts more accessible, engaging, and meaningful for students across all disciplines. In this chapter, we will unearth the strategies and best practices that can assist educators in teaching abstract ideas.

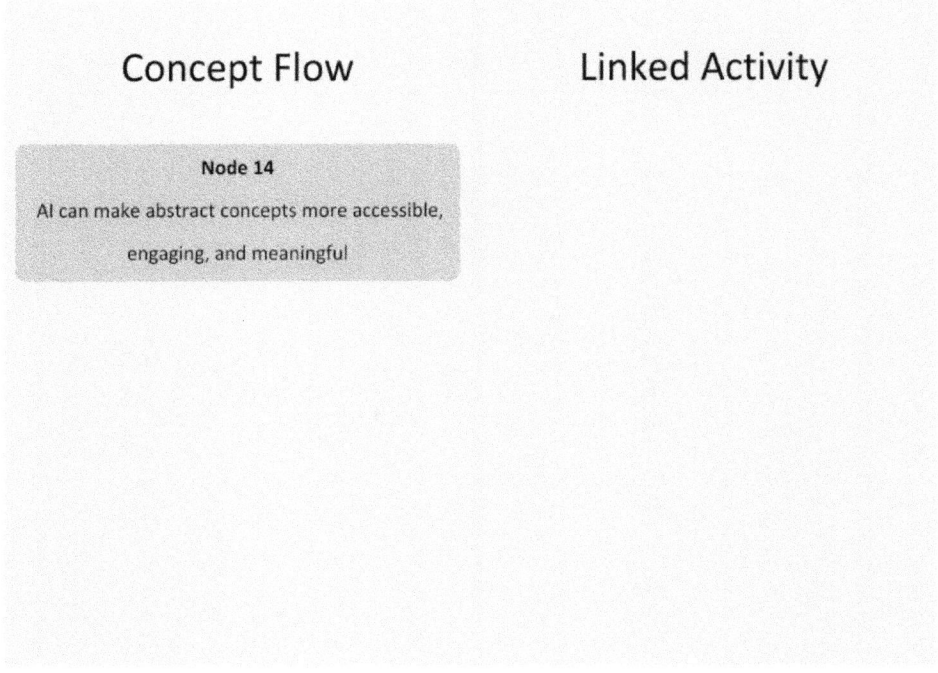

AI IN STEM EDUCATION

Virtual Labs: Immersive and Interactive Platforms

Virtual labs revolutionize STEM education by offering immersive platforms that leverage multimedia learning theory and cognitive load management (See sidebar) to create engaging experiences.

Presenting information through visual, auditory, and interactive modes helps students better process and integrate new knowledge. Immersive simulations encourage deeper processing, known as germane cognitive load, reinforcing conceptual understanding. A study on Labster's Fermentation Virtual Lab showed undergraduate students loved the opportunity to use technology to build foundational knowledge through concrete experiences, aligning with Kolb's experiential learning cycle.

For example, a virtual lab on DNA replication might begin by providing a brief overview of DNA and its importance as the genetic blueprint for life. Students are then guided through an interactive 3D model of a DNA double helix, allowing them to zoom in and observe the structure of the nitrogenous bases and the sugar-phosphate backbone. This hands-on exploration helps students visualize the abstract concept of DNA structure in a tangible way."

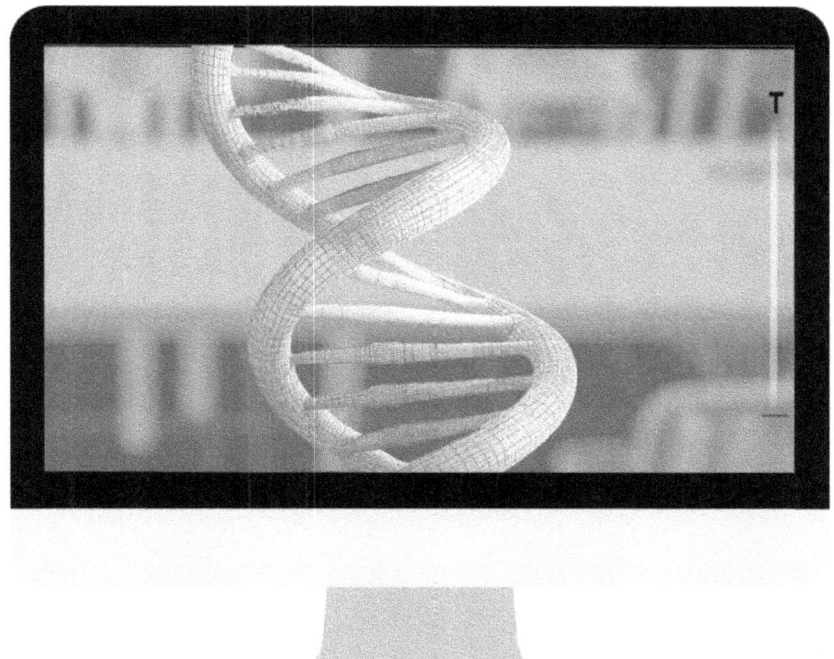

The policy mandate for K-12 science teachers to include engineering and mathematical thinking in science classrooms emphasizes the importance of virtual labs in STEM education. These tools enable students to apply age-appropriate math and science concepts to solve engineering problems, aligning with integrated STEM education. Educators can use virtual labs to create meaningful learning experiences and equip students with 21st-century skills.

AI Tools for Practical Application in STEM

AI tools reshape STEM education by serving as lab partners and guiding students through the scientific process. These tutoring systems provide real-time feedback, support, and personalized guidance, enabling students to take control of their learning journey. Open-ended scientific inquiry with AI tutors supports critical skills necessary for STEM accomplishment, such as data analysis, experimental design, collaboration, and communication.

Labster offers a virtual lab on enzyme kinetics, letting students change variables like substrate concentration and temperature to study their effects on enzyme activity. The AI tutor guides students through the experimental design process, helping them develop data analysis skills by interpreting the results and drawing conclusions. This experience sharpens students' scientific reasoning and prepares them for authentic research challenges.

Intelligent tutoring systems guiding students through the Solution-based Design Process (SBDP) show how AI supports problem-finding and critical thinking skills. These AI tools help students connect STEM content and real-world issues, encouraging them to think about interdisciplinary STEM.

Virtual labs and AI tutors offer benefits but shouldn't replace hands-on experiences and human mentorship. Educators must balance technology use with developing critical thinking and problem-solving skills through direct interaction with physical materials and equipment.

Educators and policymakers must carefully consider the ethical implications of AI in education, such as data privacy, algorithmic bias, and transparent and responsible implementation.

Despite the challenges, virtual labs and AI tools in STEM education have significant potential. As education evolves, educators must stay informed about the latest educational technology and adapt teaching practices. By integrating these tools into our curriculum, we can empower students to succeed in a complex and technology-driven world.

Balancing Virtual and Hands-On Experiences

While virtual labs simplify abstract STEM concepts, hands-on experiences and physical experiments are crucial for developing practical skills and reinforcing scientific principles. Educators should balance virtual and hands-on learning to create a well-rounded STEM curriculum. One approach is using virtual simulations as pre-lab exercises before applying the knowledge in physical labs. This boosts confidence, improves performance, and optimizes limited resources.

Next, the DNA replication lab transitions into an animation that visualizes the process step-by-step. The double helix unwinds, and enzymes like helicase and primase are introduced as key players in the replication process. Students observe how DNA polymerase assembles new complementary strands by matching free nucleotides to the exposed bases on the original strands.

As replication continues, more polymerase enzymes are recruited, forming a replication fork. Finally, any remaining gaps are filled, and the new strands are proofread for errors. Students can pause the animation at each step, zoom in on specific molecules, and access supplementary information and quizzes to reinforce their understanding.

Virtual labs can be used for complex or dangerous experiments, such as simulating nuclear reactions or studying high-velocity collisions. Hands-on activities can be reserved for essential experiments requiring physical manipulation and observation, like dissecting a specimen or building a simple machine. This approach ensures students develop conceptual understanding and practical skills.

Another effective strategy is to have students design experiments in a virtual environment and execute them in a physical lab. By comparing results from both settings, students can develop a nuanced understanding of the scientific process and the limitations of virtual simulations. This approach encourages critical thinking about factors influencing experimental outcomes and problem-solving skills while troubleshooting discrepancies.

To balance virtual and hands-on learning, educators can follow these guidelines:

- Define learning objectives for each activity and select the most appropriate format (virtual or hands-on) to achieve those goals.
- Regularly assess students' understanding and skills to ensure virtual and hands-on activities contribute to their learning.
- Seek student feedback to gauge engagement and preferences, adjusting the balance of virtual and hands-on activities.
- Stay updated with the latest research and best practices in STEM education for curriculum design and teaching strategies.

Due to limited resources or time, educators may face challenges balancing virtual and hands-on learning. To address these, educators can explore cost-effective alternatives, like using open-source virtual labs or collaborating with other institutions to share resources. They can prioritize critical hands-on activities for practical skills and conceptual understanding while using virtual labs to supplement learning.

As STEM education evolves, educators must embrace virtual labs and AI tools while recognizing the value of hands-on experiences. Integrating these approaches creates a dynamic learning environment that prepares students for the 21st century. Ongoing professional development is crucial for educators to stay current with emerging technologies and best practices.

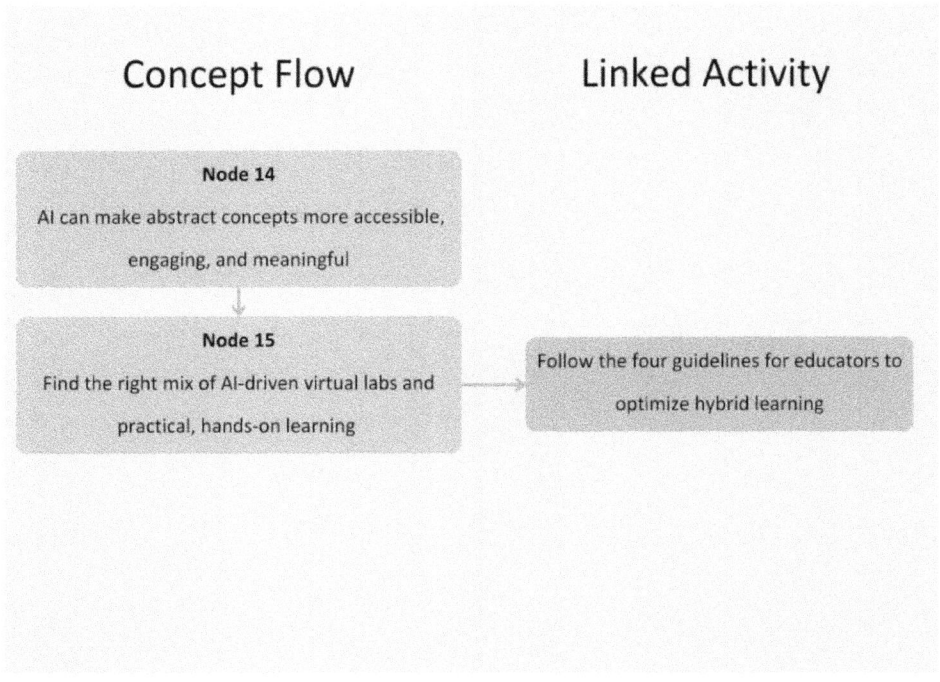

SIDEBAR: Key Terms

Virtual Labs are digital learning environments that allow students to conduct experiments, explore concepts, and develop skills without a physical laboratory. These simulations provide a safe, accessible, and cost-effective way for learners to engage with STEM subjects. Benefits include the ability to repeat experiments, manipulate variables, and visualize complex processes, making them valuable tools for enhancing STEM education.

Multimedia Learning Theory emphasizes presenting information through multiple modalities (words, images, and sounds) to facilitate effective learning. The approach is based on the idea that learners process information through different channels, and when engaged simultaneously, it leads to better retention and understanding. Virtual labs leverage multimedia learning theory by incorporating visual, auditory, and interactive elements, creating a rich, multisensory learning experience for different learning styles and preferences.

Cognitive Load refers to the mental effort needed to process new information. When it's too high, it hinders learning and causes frustration. Virtual labs manage the cognitive load by breaking down complex concepts and providing interactive simulations for self-paced learning. In a virtual lab on cell biology, cell division can be broken down into stages like interphase, prophase, metaphase, anaphase, and telophase. Each stage could be presented as a separate interactive module, allowing learners to focus on one aspect at a time.

Modalities (Visual, Auditory, Interactive) are how information is presented and processed. In virtual labs, visual modalities include images, videos, and animations to visualize complex concepts. Auditory modalities, like narration or sound effects, provide context and support, and interactive modalities allow manipulation of variables, experiments, and immediate feedback. Combining these creates an engaging, multisensory learning experience that enhances understanding and retention.

Germane Cognitive Load refers to the mental effort needed to construct and integrate new knowledge into existing cognitive schemas. Unlike extraneous cognitive load, which hinders learning, germane cognitive load promotes deep processing and conceptual understanding. Virtual labs foster germane cognitive load by providing immersive, interactive experiences that challenge learners to engage with the content, solve problems, and apply their knowledge to new situations. In a virtual lab focused on electric circuits, learners might design and build a circuit to power a small device, requiring them to apply their understanding of electrical components, circuit diagrams, and Ohm's law.

Kolb's Experiential Learning Cycle is a four-stage model that describes how people learn through experience:

1. Concrete experience
2. Reflective observation
3. Abstract conceptualization
4. Active experimentation

Virtual labs align well with this model, providing learners with hands-on, interactive experiences (concrete experience), reflection on results (reflective observation), hypothesis formation and conclusion drawing (abstract conceptualization), and testing understanding in new scenarios (active experimentation). By facilitating this cyclical learning process, virtual labs promote deep, experiential learning.

Integrated STEM education combines two or more STEM disciplines in a cohesive learning experience, emphasizing interconnectedness and real-world applications. Virtual labs support integrated STEM education by incorporating math and science concepts into engineering problem-solving scenarios. For instance, a virtual lab focused on bridge design might require learners to apply principles of physics, geometry, and materials science to create a stable structure. By engaging learners in multidisciplinary challenges, virtual labs help develop a holistic understanding of STEM subjects and their relevance in solving real-world problems.

AI-POWERED LANGUAGE LEARNING: ADAPTIVE AND IMMERSIVE EXPERIENCES

AI-powered language learning apps are transforming how people acquire new languages, offering engaging, adaptive, and effective learning experiences. These apps harness the power of AI to personalize lessons, provide instant feedback, and create immersive practice opportunities, making language learning more accessible and enjoyable for learners worldwide. By leveraging AI technologies, these platforms have the potential to foster global communication, cultural understanding, and collaboration among learners from diverse backgrounds.

Language Learning Apps: Duolingo, Rosetta Stone, and more

Popular AI-powered language learning apps like Duolingo, Rosetta Stone, and Babbel utilize advanced algorithms to offer a variety of interactive exercises and activities, such as bite-sized lessons, gamification elements, and multimedia content, to keep learners motivated and engaged. These platforms' flexibility and convenience allow users to learn at their own pace, anytime and anywhere, making language acquisition more achievable for busy learners.

Personalized learning and instant feedback

One of the most significant benefits of AI-powered language learning apps is their ability to provide personalized learning experiences and instant feedback, a crucial aspect of language learning. By analyzing user data, AI algorithms can identify each learner's strengths, weaknesses, and progress, adapting content accordingly. These tailored approaches ensure that learners receive the most relevant and effective instruction, optimizing their language acquisition journey. AI-powered apps also provide immediate corrections and guidance, helping learners improve their skills more efficiently. When a user mispronounces a word or makes a grammatical error, the app can provide real-time feedback and suggestions for improvement, accelerating the learning process and helping users develop their language skills more effectively.

Enhancing language acquisition through AI-driven practice

AI-powered language learning apps employ evidence-based techniques like spaced repetition and adaptive questioning to enhance language acquisition and retention. Spaced repetition optimizes the timing of content review based on a learner's performance, ensuring that they revisit the material at the most effective intervals. Adaptive questioning adjusts the difficulty level of exercises based on the user's responses, providing a consistently challenging and engaging learning experience.

AI-driven practice enables learners to develop their speaking, listening, reading, and writing skills in an immersive and interactive way. Advanced features like speech recognition and natural language processing allow users to practice their language skills in realistic contexts. For instance, some apps offer AI conversation partners that engage learners in lifelike dialogues, providing opportunities for authentic language use and instant feedback on pronunciation and fluency.

The role of educators in guiding students

While AI-powered language learning apps offer numerous benefits, the role of educators in guiding students to use these tools and develop self-regulated learning skills effectively remains crucial. Teachers can help students navigate these apps, set learning goals, and monitor their progress to maximize the benefits of AI-powered language learning. Educators can support and enhance student learning by incorporating these apps into their language teaching practices, creating a more engaging and effective language acquisition experience.

The integration of AI in language learning apps represents a significant step forward in making language acquisition more accessible, personalized, and effective. As AI advances, we can anticipate the development of even more innovative and powerful tools that will further revolutionize how we learn and teach languages. By harnessing the potential of AI-powered language learning apps, we can foster a global community of language learners, promoting cross-cultural understanding and collaboration.

BRINGING HISTORY AND SOCIAL SCIENCES TO LIFE WITH AI

AI is transforming history and social sciences education by enabling richly immersive and interactive learning experiences that bring abstract concepts vividly to life. By leveraging cutting-edge technologies like virtual reality (VR), augmented reality (AR), and game-based learning environments infused with AI, educators can transport students into the midst of ancient civilizations, historical events, and authentic social simulations.

VR, for example, can recreate ancient worlds in meticulous detail, allowing students to tour the grand Colosseum in Rome or explore the ingenious Incan architecture of Machu Picchu virtually. These embodied experiences, enriched by AI characters serving as guides or interlocutors, spark powerful connections to people, places, and pivotal moments.

Such environments provide fertile ground for AI-infused collaborative inquiry and game-based learning approaches. Students could work together investigating key historical or social questions, gathering evidence from AI-generated multimedia resources and datasets. Adaptive tutoring systems can personalize the journey, tailoring learning pathways based on group and individual needs. Throughout, seamless AI feedback and formative assessments integrate organically into the interactive narrative flow.

These dynamic, student-centered experiences allow learners to practice vital skills, explore cause-and-effect relationships, weigh different viewpoints, and formulate evidence-based historical arguments. The engaging simulations promote deeper conceptual understanding compared to traditional lectures or texts. Simultaneously, they cultivate empathy and social-emotional capacities through humanistic, art-based education.

Explore Ancient Civilizations and Historical Events

Imagine students investigating the Roanoke mystery through an interactive multimedia research platform powered by a custom ChatGPT model designed by their teacher. The platform could present a digital recreation of Roanoke using images, maps, artifacts, and primary source documents curated and included in the model's training data. As students explore this digital environment, they could query the custom ChatGPT at any point for relevant information, analysis, or guidance based on the specific prompts and context provided during training. For example, if they found a particular artifact like an engraved stone, they could upload an image that would trigger the AI to provide historical context, propose hypotheses about its meaning, and suggest other related lines of inquiry to pursue. The AI could even take on the persona of figures like John White based on the training data, offering roleplay dialogue to bring the narrative to life. Throughout their investigation, the custom ChatGPT could continuously evaluate their findings, identify gaps or contradictions, and pose thought-provoking questions to deepen their analysis, essentially serving as an intelligent, interactive research assistant tailored specifically for this learning experience.

The key benefit of this approach is that it allows teachers to precisely sculpt the knowledge landscape the AI draws upon and how it guides students while still providing open-ended opportunities for exploratory learning. By leveraging AI-infused collaborative inquiry and game-based pedagogies, educators can create powerful learning experiences that facilitate the active construction of historical understanding. Students work together to investigate key questions, gathering evidence from AI-curated data sources and multimedia visualizations. At the same time, adaptive AI tutors personalize learning pathways based on individual discoveries and knowledge gaps.

While profoundly engaging, these AI-powered learning experiences should align with ethical principles and sound learning objectives. As Fiske et al. (2019) and Chai et al. (2022) emphasize, the classroom integration of AI necessitates careful consideration of implications around bias, transparency, privacy, and the human role. Judiciously designed and implemented, however, AI can unlock new frontiers for bringing the richness of human experiences across eons into focus for K-12 learners, fostering critical thinking, historical empathy, and a deep understanding of the past.

Fostering Empathy, Critical Thinking, and Historical Understanding

The Roanoke investigation example shows how AI experiences can foster empathy, critical thinking, and historical understanding. As students explore the mystery from various perspectives, the AI guide presents diverse viewpoints and individual stories, challenging learners to consider the human impact of historical events. The AI shares accounts from colonists, Native Americans, and fictional characters, each offering a unique lens on the Roanoke story. Engaging with these varied perspectives helps students develop a nuanced understanding of the past and cultivate respect and compassion for those who lived through it.

As the investigation progresses, the AI tutor presents increasingly complex scenarios and datasets for analysis. The AI introduces conflicting evidence or accounts, prompting students to evaluate the credibility of sources and reconcile contradictory information. The AI poses thought-provoking questions to guide their thinking, such as "How might the colonists' motivations have influenced their actions?" or "What social and political factors contributed to the tensions between the colonists and Native Americans?" Grappling with these questions sharpens students' critical thinking skills and helps them navigate the ambiguities of historical interpretation.

Students gain a realistic understanding of AI's capabilities and limitations through these interactions. As one student noted after participating in a similar AI-enhanced learning experience,

I thought AI could do everything. Now I know AI can be stubborn and biased compared to humans.

— (DAI ET AL., 2023B)

This realization is essential for developing a critical perspective on AI and its role in shaping our understanding of the world.

AI IN ARTS AND MUSIC EDUCATION: FOSTERING CREATIVITY AND INNOVATION

Music classrooms where students actively experiment with complex concepts like harmony and composition using AI-powered tools are becoming a reality. AI in arts and music education has the potential to make abstract concepts more accessible, engaging, and meaningful for learners of all ages and skill levels.

AI-powered color palette generators like Adobe Color and Khroma can help students understand color theory by creating harmonious color schemes based on a chosen image or set of colors. These tools make the abstract concept of color harmony more tangible and accessible for learners.

Lila, a high school student, is designing a poster for her school's upcoming Earth Day celebration. She aims to create an eye-catching color scheme that effectively communicates the event's environmental theme.

Lila uses an AI-powered color palette generator like Adobe Color. She uploads an image of a lush green forest, which serves as her inspiration for the poster's color scheme.

The AI tool analyzes the image and suggests several color palettes that harmoniously capture the essence of the forest. Lila explores the different options, considering how each palette evokes specific emotions and aligns with her poster's message.

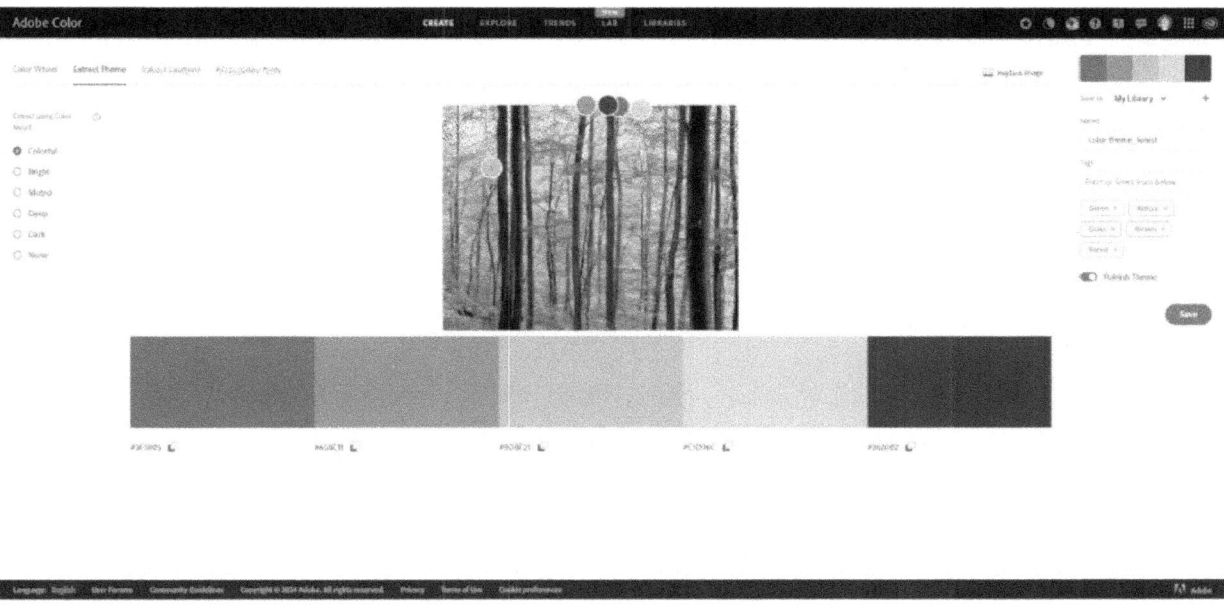

She chooses a palette featuring shades of green, brown, and blue, creating a sense of natural balance and tranquility. The AI tool provides the exact hex codes for each color, making it easy for Lila to apply the palette to her design.

As Lila works on her poster, she experiments with different color combinations and placements, using the AI-generated palette as a guide. She learns how the colors interact with each other and creates visual hierarchy and contrast through strategic color choices.

The AI color palette generator simplifies the process of selecting colors and helps Lila grasp the fundamental principles of color theory through hands-on practice. By engaging with the tool and seeing the immediate results of her choices, Lila develops a deeper understanding of the power of color in design.

AI-Assisted Composition and Style Transfer

Consider a middle school student who is passionate about music but struggling to understand how different elements create a cohesive piece. AI tools like Amper and Jukedeck allow this student to explore composition by creating their own melodies and receiving suggestions for complementary harmonies or rhythms. Moreover, AI helps students understand different musical styles through style transfer experimentation. A high school student learning about classical music could use CoCreative AI to transform their original melody into the style of Mozart or Beethoven. These tools break down complex concepts into manageable steps, enabling students to learn by doing and develop their musical intuition.

AI-assisted drawing tools like AutoDraw and Sketch RNN can help students grasp the basics of shape, form, and composition by providing suggestions and completing partial sketches. These tools break down complex drawing concepts into manageable steps, allowing students to build their skills and confidence.

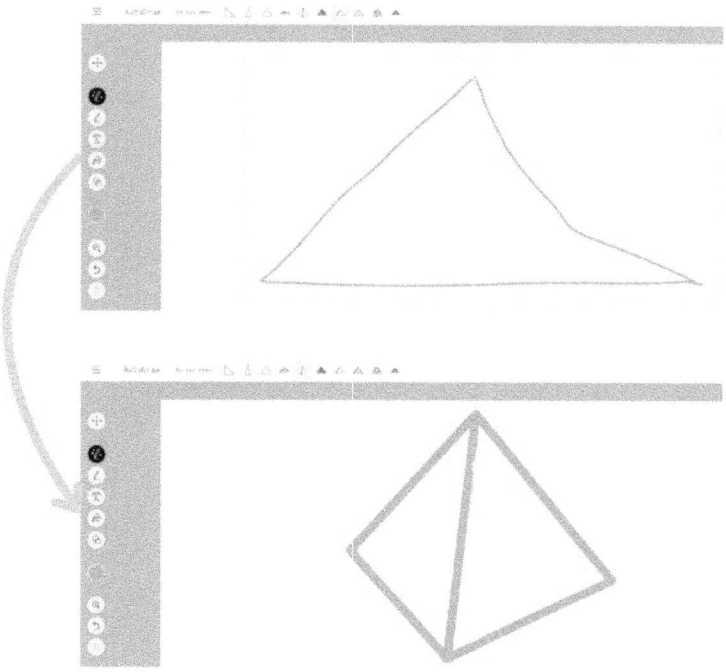

Style transfer AI tools like DeepArt and Prisma can help students understand the characteristics of different artistic styles, such as Impressionism or Cubism, by applying these styles to their own artwork. This hands-on approach makes learning about art history and styles more engaging and interactive.

Enhancing music theory understanding and creative experimentation

Music theory concepts like chord progressions and song structure can be challenging for students to grasp. However, AI tools simplify these abstract ideas, making them more approachable. Applications like Harmony Assistant and Capo use AI to analyze songs and visually represent their harmonic structure, helping students understand how chords function within a piece. Furthermore, AI offers instant feedback on students' compositions, identifying areas for improvement and providing suggestions based on music theory principles. This interactive learning process allows students to develop their understanding of music theory through creative experimentation and real-time guidance.

AI-powered image analysis tools like Google Arts & Culture and Smartify can help students understand the elements and principles of art by identifying and explaining key features in famous artworks. These tools provide instant feedback and guidance, making abstract art concepts more concrete and relatable for learners.

Balancing AI tools with human artistry and expression

While AI offers powerful tools for learning and creating, balancing these technologies with the development of human artistry and personal expression is essential. Music educators should use AI to enhance, rather than replace, students' creativity and musicality.

One way to achieve this balance is by using AI-generated elements as starting points for student compositions. For instance, a teacher could provide an AI-created melody as a prompt, asking students to build upon and personalize it using their own artistic vision. This approach encourages students to engage with AI tools creatively while still developing their unique musical voices.

Generative AI models like DALL-E, Ideogram, and Midjourney can be used to create visual prompts for student art projects, encouraging creative problem-solving and divergent thinking. By providing a starting point grounded in abstract concepts like symbolism or metaphor, these tools challenge students to think critically and create meaningful artwork.

Incorporating AI in arts and music education has the potential to transform how students learn and create. By breaking down complex concepts and providing interactive tools for experimentation, AI makes abstract ideas more accessible and personally meaningful. However, using these technologies in a way that nurtures, rather than hinders, human creativity and expression is critical. With thoughtful integration, AI can become a valuable asset in cultivating the next generation of innovative and passionate artists and musicians.

APPENDIX: HOW TO SUPPORT STUDENTS' LANGUAGE LEARNING WITH MOBILE APPS: A GUIDE FOR EDUCATORS

Key Considerations

1. Recognize the importance of students' attitudes and acceptance.
   - Positive student attitudes and acceptance of mobile apps are essential for realizing learning benefits.
   - Assess students' openness to using language learning apps and address any concerns upfront.
2. Prioritize apps with high-quality features.
   - Look for apps with features that positively influence students' acceptance, such as accuracy, completeness, relevance, and efficiency of learning content.
   - Evaluate apps for usability, perceived ease of use, and usefulness, as these factors impact students' intention to use them.
3. Ensure apps are compatible with students' devices and learning preferences.
   - Select apps that are compatible with the devices students have access to and the operating systems they use.

  - ○ Consider apps that cater to different learning styles, such as kinesthetic learners, to maximize engagement.
4. Leverage students' prior experience with mobile devices.
  - ○ Assess students' familiarity and prior experience with mobile devices, as this can influence their acceptance of language learning apps.
  - ○ Provide additional support and training for students with less mobile device experience.

Implementation Strategies

1. Introduce apps gradually and provide guidance on effective use.
  - ○ Avoid overwhelming students by assigning too many apps at once. Introduce them incrementally.
  - ○ Demonstrate how to use key app features and provide guidelines on when and how to use them for language learning.
2. Design tasks that encourage meaningful interaction with apps.
  - ○ Create assignments that require students to engage with app content in context, such as recording conversations with AI chatbots.
  - ○ Have students reflect on their app learning experiences and share with classmates how the apps are supporting their language development.
3. Monitor app engagement and gather regular feedback
  - ○ App analytics or assessments are used to track student progress and evaluate the effectiveness of the language apps being used.
  - ○ Solicit student feedback frequently to identify areas for improvement in your mobile app integration approach.
4. Highlight the positive impact on learning outcomes.
  - ○ Share research with students showing how mobile language apps can improve academic performance and engagement.
  - ○ Celebrate student progress and achievements accomplished through mobile app use to reinforce benefits and motivation.

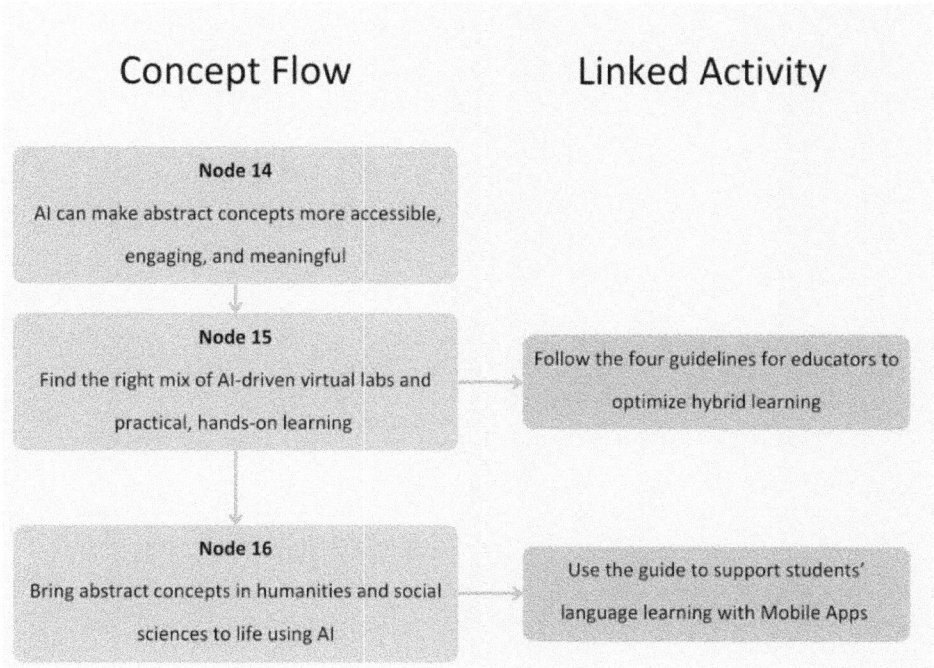

CONCLUSION

By leveraging research insights on the factors influencing students' acceptance of mobile language learning apps, educators can create a strategic, evidence-based approach to integrating these powerful tools into their instructional practice. Through careful app selection, incremental introduction, targeted assignments, and ongoing evaluation, language educators can harness the potential of mobile technologies to boost students' engagement, performance, and, ultimately, their language proficiency. Use the key considerations as a checklist for evaluating mobile language apps and implement the research-backed strategies to set your students up for mobile-assisted language learning success.

8

TRANSFORMING STEM EDUCATION

A STEP-BY-STEP GUIDE TO CREATING AI-SUPPORTED COLLABORATIVE PROBLEM-SOLVING ACTIVITIES THAT WORK

As an educator in the STEM field, you're always looking for innovative ways to engage your students and help them develop critical problem-solving skills. With the rise of artificial intelligence (AI) in education, you now have a powerful tool at your disposal to create collaborative problem-solving activities that truly kick ass. By incorporating insights from leading researchers in the field, such as Kim et al. (2022), Laat & Joksimovic (2020), Fan et al. (2021), Lester (2024), and Nixon (2024), you can design AI-supported activities that foster creativity, critical thinking, and teamwork among your students.

STEP 0: K-12 EDUCATOR SELF-ASSESSMENT RUBRIC FOR IMPLEMENTING AI-SUPPORTED COLLABORATIVE PROBLEM-SOLVING IN STEM CLASSROOMS

Use the self-assessment rubric in the appendix to evaluate your readiness to implement AI-supported collaborative problem-solving activities in your classroom. This rubric assesses your knowledge of AI tools, understanding of collaborative problem-solving principles, and ability to design engaging activities that align with your curriculum. By honestly assessing your strengths and areas for improvement, you can identify the resources and support you need to successfully integrate AI into your teaching practice and create a dynamic, student-centered learning environment that prepares your students for the challenges of the 21st century

STEP 1: IDENTIFY THE PROBLEM-SOLVING SKILLS YOU WANT TO TARGET

As you plan your AI-supported collaborative problem-solving activity, it's crucial to align the targeted skills with your learning objectives, ensuring relevance and meaning for your students (Kim et al., 2022). By focusing on skills that directly relate to your curriculum, you demonstrate the practical value of the activity, even to the most skeptical audience members.

Consider targeting the following problem-solving skills:

1. Critical thinking
2. Creativity
3. Data analysis
4. Logical reasoning
5. Adaptability
6. Collaboration
7. Communication
8. Leadership
9. Digital literacy

Next, consider your learners' current status and preferences when selecting appropriate learning content and problem-solving skills (Kim & Kim, 2020). Tailoring the activity to your students' needs and interests can increase engagement and buy-in, even from those hesitant about AI in education.

Prioritize complex problem-solving skills that are particularly relevant in today's workplace, as they align with advanced technological innovations and facilitate creative thinking for dealing with dynamic, interdisciplinary, and complex situations (Laat & Joksimovic, 2020). By emphasizing these skills' real-world applications, you can demonstrate the activity's long-term value to critical stakeholders.

Existing AI curriculum designs, such as the two-year high school AI curriculum proposed by Bellas et al. (2022), can serve as models for integrating AI education and problem- solving skills into your educational framework. Leveraging established best practices and proven models can build credibility and address potential concerns from skeptical audience members.

Involve teachers in the process of collaboratively designing integrated AI curricula for K-12 classrooms to ensure that the targeted problem-solving skills align with pedagogical goals and student needs (Lin, 2020). Engaging teachers in the design process can foster buy-in and demonstrate that the activity is grounded in sound educational principles, even to the most critical observers.

Finally, it addresses psychological barriers faced by underrepresented groups in STEM, such as stereotype threats, lack of belonging, and hostile environments, by targeting problem-solving skills that promote inclusivity and diversity (Nixon, 2024). Prioritizing equity and inclusion can demonstrate the activity's social and ethical value, even to those skeptical of AI's potential benefits.

Link to example: The climate change investigation activity targets critical thinking, data analysis, teamwork, and communication skills, which are crucial problem-solving skills in STEM. By focusing on these skills, the activity aligns with the learning objectives of the 10th-grade science curriculum and prepares students for real-world challenges in environmental science.

STEP 2: SELECT AN APPROPRIATE AI TOOL OR PLATFORM

With a clear understanding of the problem-solving skills you want to target, it's time to research and evaluate AI tools that can support collaborative problem-solving in your classroom (Laat & Joksimovic, 2020). Look for tools that offer:

1. Ease of use and intuitive interfaces
2. Adaptability to different learning styles and needs
3. Real-time feedback and support for students
4. Integration with existing educational platforms

To assess the appropriateness of an AI tool, start by evaluating its ease of use and adaptability (Trovato & Russo, 2021). Consider how well the tool can be integrated into your existing educational environment and workload (Elzain et al., 2022). Identify potential barriers to implementation, such as technical requirements or training needs (Mlodzinski et al., 2023).

Next, examine the impact of the AI tool on academic achievement and problem-solving skills in your specific subject area (Nağaç & Kalayci, 2021). Look for evidence of its effectiveness in enhancing learning outcomes and addressing individual student needs through personalization (Murtaza et al., 2022).

Assess your own technological and pedagogical content knowledge levels (Emre & Çelik, 2021) to determine if you have the necessary skills and understanding to effectively utilize the AI tool in your teaching practices.

Additionally, consider the acceptability and usability of the AI tool among your target student population (Schilling et al., 2023). Evaluate the tool's user interface, accessibility features, and overall user experience to ensure it aligns with your students' needs and abilities.

Finally, investigate the psychometric properties of the AI tool, particularly if it will be used for student assessment (Calatayud et al., 2021). Ensure that the tool provides reliable and valid measures of student performance and aligns with your assessment goals.

Some popular AI tools for education include:

1. Cognii
2. Century Tech
3. Thinkster Math (Fan et al., 2021)
4. Individualized AI tutor that integrates three developmental learning networks (DLNs) based on a deep adaptive resonance theory (Deep ART) network (Kim & Kim, 2020)
5. Multi-modal guided environment to assist children in learning to solve a Rubik's Cube with automatic solving and interactive explanations (Kausik et al., 2022)

When selecting an AI tool, consider factors such as ease of use, adaptability, and the ability to provide real-time feedback and support (Laat & Joksimovic, 2020). Educators need to have an understanding of various AI applications in education, such as software that adjusts learning materials based on students' needs and tools that evaluate teaching performance, to effectively utilize AI tools in enhancing teaching and learning experiences (Huang et al., 2021).

Look for AI tools that can capture fine-grained learner behavior, enabling the development of analytics that model diverse learners at scale and recognize that different aspects of a person's identity can intersect and interact with each other (Nixon, 2024). This allows for identifying areas where different subgroups of learners require support.

The "AI Tool Suitability Rubric for Educators" in the appendix can help determine if an AI tool is appropriate for a specific subject, grade level, and student population. However, this rubric is only a guide and should be used alongside other factors like cost and practicality.

Link to example: The climate change investigation activity incorporates AI data analysis software and digital collaboration tools, which are suitable for the project's objectives and the students' grade level. These tools enable students to gather, analyze, and interpret environmental data effectively, while also promoting teamwork and communication skills.

STEP 3: DESIGN THE COLLABORATIVE PROBLEM-SOLVING ACTIVITY

In this step, focus on designing an engaging collaborative problem-solving activity that seamlessly integrates the selected AI tool. Begin by creating an authentic scenario or challenge that requires students to work together and apply their problem-solving skills (Lester, 2024). The EngageAI Institute's AI-driven narrative-centered learning environments, which generate interactive story-based problem scenarios, can serve as inspiration for creating dynamic and tailored learning experiences (Lester, 2024).

Incorporate the AI tool as a supportive "team member" that offers insights, suggestions, and guidance throughout the problem-solving process (Nixon, 2024). The EngageAI Institute's embodied conversational agents, with their multiple modalities, can take on roles such as virtual mentors, learning companions, collaborators, and facilitators (Lester, 2024). These AI teammates can provide relevant data, suggest alternative solutions, offer feedback on student ideas, and facilitate communication and collaboration among team members.

The importance of leveraging AI that can generate new content as "teammates" in human-AI collaboration to mitigate biases, support underrepresented group members, and promote equitable participation in STEM teams is highlighted by Nixon (2024).

Memmert & Bittner (2022) provide valuable guidance for designing suitable research contexts through their literature review on research contexts for studying complex problem-solving through human-AI collaboration.

To enhance student engagement, motivation, and 21st-century skills, consider incorporating online game-based learning elements into the activity (Zahra et al., 2022). Develop clear instructions and guidelines for student collaboration and interaction with the AI tool to ensure a smooth and productive experience. The learner status DLN can help the AI tutor adapt to new types of information about the learner observed during the learning process (Kim and Kim, 2020).

When determining the specific problem-solving skills to focus on, provide a clear rationale for why these skills are essential in the context of AI-supported collaborative learning

(Kim et al., 2022). Refer to established AI education frameworks, such as the "AI for K-12" curriculum framework by the AI4K12 Initiative, to demonstrate the alignment of targeted skills with recognized standards (Touretzky et al., 2019).

By designing an engaging collaborative problem-solving activity that seamlessly integrates AI tools as supportive "team members," educators can create dynamic and tailored learning experiences that foster the development of essential problem-solving skills in an AI-supported environment.

Link to Example: The climate change investigation example demonstrates how to design a collaborative problem-solving activity that seamlessly integrates AI tools as supportive "team members." In this 4-week project, students work together to investigate the impact of climate change on their local community. The AI tool provides relevant background information, suggests research methods, facilitates communication, and offers guidance throughout the project.

STEP 4: MONITOR AND SUPPORT THE COLLABORATIVE PROBLEM-SOLVING PROCESS

As students engage in AI-supported collaborative problem-solving activity, it's essential for educators to monitor their progress and provide support when needed. Regularly check in with each team to assess their understanding, collaboration dynamics, and interaction with the AI tool.

AI can be used to monitor student learning in real-time, providing valuable insights for educators to offer targeted support (Huang et al., 2021). AI-powered learning analytics can track student progress, identify areas of difficulty, and suggest appropriate interventions (Karaca et al., 2021).

Encourage students to ask questions, share their thought processes, and reflect on their experience working with the AI tool (Kim et al., 2022). Prompt them to consider how the AI is supporting their collaboration and problem-solving efforts, and discuss any challenges they may be facing.

Be prepared to offer guidance and clarification as needed, particularly if students encounter technical issues or have difficulty interpreting the AI tool's suggestions. Provide additional resources or mini-lessons to reinforce key concepts and skills related to the problem-solving activity (Laat & Joksimovic, 2020).

Foster a supportive and inclusive learning environment that values diverse perspectives and encourages equal participation (Nixon, 2024). Monitor team dynamics to ensure that all students have the opportunity to contribute and that the AI tool is being used in a way that promotes equity and fairness.

The importance of trust in educators' attitudes towards AI-based educational technology is highlighted by the active monitoring and support of students' use of the AI tool, which can build trust in the technology and demonstrate its value in enhancing the learning experience (Nazaretsky et al., 2021).

Throughout the monitoring process, educators should also reflect on their own practices and the effectiveness of the AI-supported activity (Rushton et al., 2022). Consider how the activity can be refined or adapted for future implementation based on student feedback and observed outcomes.

Link to example: In the climate change investigation example, the teacher regularly checks in with each team to monitor their progress and provide support. The teacher uses AI-powered analysis of student data to track student progress, identify areas where teams may be struggling, and offer targeted guidance. By fostering a supportive and inclusive learning environment, the teacher ensures that all students have the opportunity to contribute and that the AI tool is being used in a way that promotes equity and fairness. This active monitoring and support help build trust in the AI technology and demonstrate its value in enhancing the collaborative problem-solving experience.

STEP 5: MONITOR AND FACILITATE THE COLLABORATIVE PROBLEM-SOLVING PROCESS

As your students work on the collaborative problem-solving activity, your role is to monitor their progress and provide guidance and support as needed. Observe how they interact with each other and the AI tool, and look for opportunities to encourage active engagement with the AI tool, foster a collaborative environment that values diverse perspectives, prompt students to build upon each other's ideas and insights, and address any challenges or roadblocks that may arise (Fan et al., 2021).

The importance of establishing a culture of collaborative learning among teachers and a "safe to fail" environment for students for effective implementation of student-AI collaboration is emphasized by Kim et al. (2022). By facilitating the collaborative problem-solving process, you can help your students maximize the benefits of working with the AI tool and develop their problem-solving skills (Fan et al., 2021). Encouraging students to actively engage with the AI tool, seeking its insights and recommendations, is recommended by Laat & Joksimovic (2020).

The EngageAI Institute is creating an innovative framework that uses a variety of data streams from students' conversations, gaze, expressions, gestures, etc. as they interact with each other, teachers, and conversational agents to analyze their learning process. This can drive real-time narrative generation and pedagogical decisions to support the collaborative problem-solving process (Lester, 2024). Adaptive, personalized AI systems incorporating support for understanding and managing social interactions and emotions can significantly enhance learning experiences in diverse STEM teams (Nixon, 2024).

The significance of reflective professional learning within collaborative communities and the importance of teachers having adequate technology competence to effectively incorporate AI tools into their teaching practices are emphasized (Rushton et al., 2022). As you guide educators in creating scenarios or challenges that require students to work together and apply their problem-solving skills, draw connections to the study, which demonstrates how game-based learning can enhance collaboration, critical thinking, and problem-solving skills in students (Zahra et al., 2022; Lester, 2024).

In the climate change investigation example, the teacher has numerous opportunities to observe and guide students throughout the data collection, analysis, and presentation phases. By encouraging active engagement with the AI tools and fostering a collaborative environment, the activity promotes effective problem-solving and knowledge sharing among students. This hands-on approach allows the teacher to provide timely support and ensure that students are maximizing the benefits of working with the AI tool to develop their problem-solving skills.

STEP 6: DEBRIEF AND REFLECT ON THE ACTIVITY

After completing the collaborative problem-solving activity, it's essential to conduct a debriefing session where students can share their experiences, challenges, and successes. Encourage them to reflect on how the AI tool supported their problem-solving process, what they learned from collaborating with the AI and their peers, and how they can apply their new problem-solving skills in other contexts.

This reflection process is crucial for reinforcing learning and helping students internalize the benefits of AI-supported collaborative problem-solving (Laat & Joksimovic, 2020).

Encouraging students to reflect on how the AI tool supported their problem-solving process and what they learned from collaborating with it is emphasized (Laat & Joksimovic, 2020).

The importance of collaboration among researchers, educators, policymakers, and technology developers for harnessing AI's power to create diverse, inclusive, and equitable STEM ecosystems is highlighted (Nixon, 2024). Debriefing sessions provide an opportunity to gather valuable feedback that can inform this collaborative effort.

During the debriefing session, encourage educators to refer to the "AI Foundations for Everyone" course by IBM (IBM, 2021) and the "AI in Education" resources by Google for Education (Google for Education, 2022). These resources can help educators and students contextualize their experiences within the broader landscape of AI in education, understanding AI fundamentals, data science, and ethical issues (Karaca et al., 2021).

Issues of trust influencing educators' attitudes towards AI-based educational technology are explored (Nazaretsky et al., 2021), emphasizing the importance of building trust in AI tools for educators to embrace and effectively utilize these resources in the classroom.

In the climate change investigation example, the class symposium and the reflection on AI's role in the activity provide valuable opportunities for students to share their experiences, challenges, and successes. By encouraging students to reflect on how the AI tools supported their problem-solving process and the implications of AI in scientific research, the activity reinforces learning and helps students internalize the benefits of AI- supported collaborative problem-solving. This debriefing session allows students to consolidate their understanding and appreciate the potential of AI in tackling complex, real-world issues like climate change.

STEP 7: ASSESS AND ITERATE

After completing the AI-supported collaborative problem-solving activity, evaluating its effectiveness is crucial based on student feedback, learning outcomes, and engagement levels.

The assessment of student-AI collaboration should focus on understanding the process students undergo, looking at their conceptual and procedural knowledge of both the subject matter and AI technology. They also highlight the importance of assessing team-level learning, including student-student and student-AI interactions, in addition to individual assessment (Kim et al., 2022).

Consider what worked well and what could be improved, how students responded to the AI tool and the collaborative nature of the activity, and whether the activity successfully targeted the desired problem-solving skills. Use this assessment to identify areas for improvement and make data-driven decisions to refine the activity for future implementation (Kim et al., 2022; Laat & Joksimovic, 2020). A learner experience DLN that is updated to immediately reflect alteration of the educational effectiveness in the current classification, using an alternative template learning algorithm to make the classifications flexible in both expanding and contracting ways based on educational effectiveness, is proposed (Kim & Kim, 2020).

The importance of assessing collaborative skills using AI-driven methods that analyze learners' natural interactions is emphasized, as this approach reduces potential biases often found in self-reported evaluations and provides a more objective evaluation of team dynamics (Nixon, 2024).

As educators evaluate the effectiveness of AI-supported collaborative problem-solving activity, they emphasize the importance of ongoing reflection, feedback, and evaluation to monitor progress, address challenges, and make necessary adjustments to the learning experience (Kim et al., 2022; Merchie et al., 2018). This will help demonstrate to your hypercritical audience members that the process is iterative and responsive to feedback.

Tailored professional development programs have been proven effective in enhancing educators' trust in AI-powered educational technology (Nazaretsky et al., 2022), and providing opportunities for educators to evaluate AI tools and become familiar with their functionalities is essential for building trust (King et al., 2022).

In the climate change investigation example, the activity includes clear assessment criteria based on research quality, presentation, collaboration, and engagement. By evaluating the effectiveness of the activity based on these criteria and student feedback, the educator can identify areas for improvement and make data-driven decisions to refine the activity for future implementation. This ensures a continuously evolving and effective learning experience that prepares students for success in the STEM field by fostering critical problem-solving skills in an AI-supported environment.

APPENDIX

The appendix provides a set of valuable tools and guides to help K-12 educators effectively evaluate and analyze the impact of AI-supported collaborative problem-solving activities in their STEM classrooms.

While these tools may seem overwhelming at first glance, they are intended to be user-friendly and adaptable to various classroom contexts, serving as a starting point for educators to customize and refine their evaluation processes based on their unique needs and goals.

These resources, including

- Guidelines for Effective Observations
- Example of Observation Checklist for AI-Supported Collaborative Problem-Solving Observation
- Example of Survey: Student Engagement Survey for AI-Supported Collaborative Problem-Solving Activities
- How to use ChatGPT for Analyzing the Student Engagement Survey
- How to use ChatGPT for Analyzing Observations in AI-Supported STEM Activities
- Example of Developed Collaborative Inquiry Activity
- AI Tool Suitability Rubric for Educators: A Comprehensive Guide
- K-12 Educator Self-Assessment Rubric for Implementing AI-Supported Collaborative Problem-Solving in STEM Classrooms

Guidelines for Effective Observations

This guide is designed to help K-12 educators evaluate the effectiveness of AI-supported collaborative problem-solving activities in STEM classrooms in terms of AI use only.

Educators can make data-driven decisions to iteratively improve their teaching practices by assessing key metrics through a survey.

To conduct effective observations and gather meaningful data, follow these simple steps:

1. Observe multiple sessions: Aim to observe at least 3 to 5 different sessions of the same AI-supported collaborative problem-solving activity. This will help you capture a wide range of student behaviors and performance levels, minimizing the impact of anomalies.
2. Watch diverse student groups: Make sure to observe various student groups and interactions. If the activity is repeated with the same students, try to observe different group compositions over time to see how changes in group dynamics influence the outcomes.

3. Use a consistent checklist: Maintain consistency in your data collection by using the same observation checklist for each session. This standardized approach will make it easier to track changes and identify trends across different sessions.

4. Cover all key areas: During each observation, record data points for all categories listed in the checklist, including student engagement, problem-solving skills, collaboration, and AI technology integration. This comprehensive coverage will allow for a more thorough analysis of the learning experience.

Example of Observation Checklist for AI-Supported Collaborative Problem-Solving Observation

Date:

Class:

Activity:

Section 1: Student Engagement

- Students actively participate in the collaborative problem-solving tasks. Rate on a scale of 1 to 5 (1 = Never, 5 = Always)
- Students demonstrate enthusiasm and interest in the AI-supported activities. Rate on a scale of 1 to 5 (1 = Not at all enthusiastic, 5 = Very enthusiastic)
- Students ask questions and seek clarification when needed. Rate on a scale of 1 to 5 (1 = Never, 5 = Always)
- Students remain focused and on-task throughout the activity. Rate on a scale of 1 to 5 (1 = Never, 5 = Always)
- Students engage in meaningful discussions with their peers. Rate on a scale of 1 to 5 (1 = Never, 5 = Always)

Section 2: Problem-Solving Skills Development

- Students apply critical thinking skills to analyze problems. Rate on a scale of 1 to 5 (1 = Not at all, 5 = Extensively)
- Students brainstorm and generate multiple solutions. Rate on a scale of 1 to 5 (1 = Never, 5 = Always)
- Students evaluate and select the most appropriate solutions. Rate on a scale of 1 to 5 (1 = Poorly, 5 = Excellently)

- Students use AI technologies effectively to support their problem-solving process. Rate on a scale of 1 to 5 (1 = Not at all effectively, 5 = Very effectively)
- Students demonstrate creativity and innovation in their problem-solving approaches. Rate on a scale of 1 to 5 (1 = Not at all creative, 5 = Highly creative)

Section 3: Collaboration and Teamwork

- Students actively listen to their peers' ideas and perspectives. Rate on a scale of 1 to 5 (1 = Never, 5 = Always)
- Students contribute their own ideas and insights to the group. Rate on a scale of 1 to 5 (1 = Never, 5 = Always)
- Students provide constructive feedback and support to their teammates. Rate on a scale of 1 to 5 (1 = Never, 5 = Always)
- Students distribute tasks and responsibilities equitably within the group. Rate on a scale of 1 to 5 (1 = Never, 5 = Always)
- Students work together to resolve conflicts and reach consensus. Rate on a scale of 1 to 5 (1 = Never, 5 = Always)

Section 4: AI Technology Integration

- Students demonstrate proficiency in using the AI technologies provided. Rate on a scale of 1 to 5 (1 = Not proficient, 5 = Highly proficient)
- Students leverage AI technologies to enhance their problem-solving strategies. Rate on a scale of 1 to 5 (1 = Not at all, 5 = Extensively)
- Students interpret and apply insights from AI-generated data or outputs. Rate on a scale of 1 to 5 (1 = Poorly, 5 = Excellently)
- Students critically evaluate the limitations and potential biases of AI technologies. Rate on a scale of 1 to 5 (1 = Never, 5 = Always)
- Students use AI technologies ethically and responsibly. Rate on a scale of 1 to 5 (1 = Not at all, 5 = Always)

Example of Survey: Student Engagement Survey for AI-Supported Collaborative Problem-Solving Activities

Instructions: Please answer the following questions honestly based on your experience with AI-supported collaborative problem-solving activities in your STEM classroom. Your responses will help us improve these activities for future learning experiences.

1. I found the AI-supported collaborative problem-solving activities engaging and interesting. a) Strongly Agree b) Agree c) Neutral d) Disagree e) Strongly Disagree

2. The AI technologies used in the collaborative activities enhanced my learning experience. a) Strongly Agree b) Agree c) Neutral d) Disagree e) Strongly Disagree

3. I actively participated in the collaborative problem-solving tasks. a) Always b) Often c) Sometimes d) Rarely e) Never

4. The AI-supported activities motivated me to learn more about the STEM concepts covered. a) Strongly Agree b) Agree c) Neutral d) Disagree e) Strongly Disagree

5. I felt confident in my ability to contribute to the collaborative problem-solving process. a) Strongly Agree b) Agree c) Neutral d) Disagree e) Strongly Disagree

6. The AI-supported activities helped me develop better problem-solving skills. a) Strongly Agree b) Agree c) Neutral d) Disagree e) Strongly Disagree

7. I enjoyed working with my peers during the AI-supported collaborative activities. a) Strongly Agree b) Agree c) Neutral d) Disagree e) Strongly Disagree

8. What aspects of the AI-supported collaborative problem-solving activities did you find most engaging or helpful for your learning?

9. How could the AI-supported collaborative problem-solving activities be improved to enhance your engagement and learning experience?

10. Any additional comments or feedback regarding your experience with AI-supported collaborative problem-solving activities in your STEM classroom?

How to use ChatGPT for Analyzing the Student Engagement Survey

This guide will help you analyze the results from the "Student Engagement Survey for AI-Supported Collaborative Problem-Solving Activities" in your STEM classroom using ChatGPT. The steps below are designed to assist you in calculating statistics, identifying trends, and deriving actionable insights directly from your survey data.

Materials Needed

- Survey results data (preferably in a spreadsheet)
- Access to ChatGPT (or a similar AI language model capable of data analysis)

Step-by-Step Guide

1. **Prepare Your Data**
   - Gather your survey results, ensuring that quantitative data from scaled questions and qualitative responses from open-ended questions are clearly organized in your spreadsheet.

2. **Analyze Quantitative Data**
   - **Prompt for Summary Statistics**: Ask ChatGPT to calculate the percentage of responses for each option (Strongly Agree to Strongly Disagree) for each question. Example prompt:
     - "Calculate the percentage of students who selected each response for the question about engagement in AI-supported activities."
   - **Prompt for Trends and Patterns**: Ask ChatGPT to identify any notable patterns in the data, such as correlation between high engagement and positive learning outcomes. Example prompt:
     - "Identify any patterns in responses between student engagement and their enjoyment of the activities."

3. **Extract Insights from Qualitative Data**
   - **Prompt for Theme Extraction**: Input the responses to open-ended questions and ask ChatGPT to summarize common themes and suggestions. Example prompt:
     - "What are the common themes from student responses on what aspects of the AI-supported activities they found most engaging?"
   - **Prompt for Improvement Suggestions**: Ask ChatGPT for actionable suggestions based on the qualitative feedback. Example prompt:
     - "Based on student feedback, what improvements can be made to enhance engagement and learning in AI-supported activities?"

4. **Generate Actionable Recommendations (Optional)**
   - **Prompt for Recommendations**: After analyzing both quantitative and qualitative data, ask ChatGPT for specific recommendations on how to improve STEM activities. Example prompt:
     - "Generate recommendations for improving student outcomes based on the analysis of survey results."

How to use ChatGPT for Analyzing Observations in AI-supported STEM Activities

This guide will help you analyze the observation data from the "AI-Supported Collaborative Problem-Solving Observation Checklist" in your STEM classroom using ChatGPT. The following steps are designed to assist you in quantifying behaviors, identifying trends, and deriving actionable insights directly from your observation data.

Materials Needed

- Observation checklist data (ideally digitized in a spreadsheet)
- Access to ChatGPT (or a similar AI language model capable of data analysis)

Step-by-Step Guide

1. **Prepare Your Data**
   - **Gather your observations**, ensuring that data from your quantitative checklist and any qualitative notes are clearly organized in your spreadsheet.
2. **Analyze Quantitative Data**
   - **Prompt for Summary Statistics**: Ask ChatGPT to calculate the average rating for each checklist item. Example prompt:
     - "Calculate the average rating for student participation in AI-supported problem-solving tasks."
   - **Prompt for Trends and Patterns**: Ask ChatGPT to identify any notable patterns in the data, such as correlations between high participation rates and enhanced problem-solving skills. Example prompt:
     - "Identify any patterns in ratings across different sessions or groups."
3. **Extract Insights from Qualitative Data**
   - **Prompt for Theme Extraction**: Input the qualitative notes about specific instances or standout behaviors and ask ChatGPT to summarize common themes or unusual observations. Example prompt:
     - "What are the common themes from observations about student engagement during AI-supported activities?"
   - **Prompt for Improvement Suggestions**: Ask ChatGPT for actionable suggestions based on qualitative observations. Example prompt:
     - "Based on detailed notes, what improvements can be made to increase engagement and effectiveness in AI-supported problem-solving activities?"

4. **Generate Actionable Recommendations**
   - **Prompt for Recommendations**: After analyzing both quantitative and qualitative data, ask ChatGPT for specific recommendations on how to improve the STEM activities. Example prompt:
     - "Generate recommendations for improving student outcomes based on the analysis of observation data."

Example: Developed Collaborative Inquiry Activity: Investigating Climate Change with AI

Grade Level: 10th Grade

Subject: Science (Environmental Science focus)

Duration: 4 weeks

Tools Needed: AI data analysis software, internet access, digital collaboration tools (like Google Docs or Microsoft Teams)

Overview:

Students will engage in a project-based learning activity that utilizes AI tools to investigate the impacts of climate change on their local environment. The project encourages teamwork, data analysis, and presentation skills.

Objectives:

- Understand the basic science of climate change.
- Learn how to collect, analyze, and interpret environmental data using AI.
- Develop teamwork and communication skills through collaborative inquiry.

Activity Breakdown:

Week 1: Introduction and Team Formation

- **Introduction to Climate Change**: Brief lectures and interactive AI-powered presentations on climate change basics, causes, effects, and current research.
- **Team Formation**: Students are divided into small groups. Each group selects a specific aspect of climate change to study (e.g., temperature changes, precipitation patterns, impact on local flora and fauna).

Week 2: Data Collection

- **Using AI to Gather Data**: Students learn how to use AI tools to gather historical weather and environmental data from online databases.
- **Field Work**: If possible, students collect current local environmental data (like temperature and air quality) using sensors or school-provided kits.

Week 3: Data Analysis

- **AI Data Analysis Workshop**: Students receive training on how to use AI software to analyze the collected data. They learn how to identify trends, patterns, and anomalies.
- **Collaborative Analysis**: Teams use AI tools to analyze their datasets, discussing their findings and hypotheses about the impact of climate change on their chosen subject.

Week 4: Presentation and Discussion

- **Preparation of Findings**: Each team uses digital tools to create a presentation of their findings, incorporating graphs, charts, and other visual aids generated by the AI software.
- **Class Symposium**: Teams present their findings to the class. This includes a discussion session where students critique and question each other's research methods and conclusions.
- **Reflection on AI's Role**: Students discuss how AI tools enhanced their research and consider the implications of AI in scientific research and everyday life.

Assessment:

- **Research Quality**: Accuracy and thoroughness of the data collected and analyzed.
- **Presentation**: Clarity, creativity, and comprehensiveness of the final presentation.
- **Collaboration and Engagement**: Active participation in all phases of the project and contribution to team discussions.

AI Tool Suitability Rubric for Educators: A Comprehensive Guide

This rubric is designed to help educators evaluate the suitability of AI tools for their specific subject, grade level, and students. By assessing various factors such as ease of use, alignment with educational environment, impact on academic achievement, personalization, and more, educators can make informed decisions about incorporating AI tools into their teaching practice. The rubric serves as a comprehensive guide to help educators navigate the complex landscape of AI in education and select tools that best fit their unique needs and context.

To use this rubric, rate the AI tool on each criterion using the provided scale. Sum the scores for all applicable criteria. A higher total score indicates a more suitable AI tool for your specific subject, grade level, and students. However, it is essential to keep in mind that this rubric serves as a guide and should be used in conjunction with other factors, such as cost and practical considerations.

For example, an AI tool may score highly on several criteria, demonstrating strong potential for enhancing student learning and engagement. However, if the tool is prohibitively expensive or requires significant resources that are not available in your educational setting, it may not be a practical choice despite its high rubric score.

Educators should use this rubric as a starting point for evaluating AI tools and consider the results alongside other crucial factors, such as budget constraints, available resources, and long-term sustainability. By combining the insights gained from this rubric with a holistic assessment of their unique educational context, educators can make well-informed decisions about integrating AI tools that are both effective and practical for their students.

1. Ease of Use and Adaptability
   - Highly adaptable and easy to integrate into existing educational environment: The AI tool seamlessly integrates with current technology infrastructure and requires minimal training for educators and students. (4 points)
   - Moderately adaptable and requires some effort to integrate: The AI tool is compatible with existing technology but may require some adjustments or additional training for educators and students. (3 points)
   - Somewhat adaptable but requires significant effort to integrate: The AI tool requires significant changes to the existing technology set-up and demands extensive training for educators and students. (2 points)

- Not adaptable and difficult to integrate into existing environment: The AI tool is incompatible with the current technology infrastructure (e.g., requires Windows but only Macs are available) and would necessitate a complete overhaul of the system. (1 point)

2. Alignment with Educational Environment and Workload
    - Seamlessly aligns with current educational environment and workload: The AI tool naturally fits into the existing curriculum and does not add significant additional work for educators. (4 points)
    - Mostly aligns with educational environment and workload, with minor adjustments needed: The AI tool largely aligns with the curriculum but may require minor modifications to lesson plans or slightly increase educator workload. (3 points)
    - Partially aligns with educational environment and workload, requiring moderate adjustments: The AI tool somewhat aligns with the curriculum but necessitates significant changes to lesson plans and noticeably increases educator workload. (2 points)
    - Does not align with educational environment and workload, requiring major adjustments: The AI tool is not compatible with the existing curriculum and would require a complete restructuring of lesson plans and substantially increase educator workload. (1 point)

3. Barriers to Implementation
    - No significant barriers to implementation: The AI tool can be easily implemented without any notable obstacles, such as cost, technical requirements, or training needs. (4 points)
    - Minor barriers that can be easily addressed: The AI tool has some small barriers to implementation, such as a short learning curve or minimal additional costs, which can be quickly overcome. (3 points)
    - Moderate barriers that require some effort to overcome: The AI tool presents several barriers, such as moderate costs, technical requirements, or training needs, which will require dedicated effort and resources to address. (2 points)
    - Significant barriers that are difficult to overcome: The AI tool has major barriers to implementation, such as high costs, extensive technical requirements, or significant training needs, which will be challenging to overcome given available resources. (1 point)

4. Impact on Academic Achievement and Problem-Solving Skills
    - Strong evidence of positive impact on academic achievement and problem-solving skills: Numerous studies or user testimonials demonstrate the AI tool's

significant positive influence on student performance and problem-solving abilities in the specific subject area. (4 points)

- Moderate evidence of positive impact on academic achievement and problem-solving skills: Some studies or user testimonials suggest the AI tool has a positive effect on student performance and problem-solving skills in the subject area. (3 points)
- Limited evidence of positive impact on academic achievement and problem-solving skills: There is minimal research or anecdotal evidence supporting the AI tool's positive impact on student achievement and problem-solving skills in the subject area. (2 points)
- No evidence of positive impact on academic achievement and problem-solving skills: There is no available research or user feedback indicating that the AI tool positively influences student performance or problem-solving skills in the subject area. (1 point)

5. Personalization and Addressing Individual Student Needs
- Highly personalized and effectively addresses individual student needs: The AI tool offers a wide range of customization options and adaptive features that cater to diverse learning styles, abilities, and backgrounds. (4 points)
- Moderately personalized and addresses most individual student needs: The AI tool provides some customization options and adaptive features that address the needs of most students but may not accommodate all learners. (3 points)
- Somewhat personalized but may not address all individual student needs: The AI tool offers limited customization options and adaptive features, which may not adequately support the diverse needs of all students. (2 points)
- Not personalized and does not address individual student needs: The AI tool does not provide any customization options or adaptive features to cater to individual student needs. (1 point)

6. Educator's Technological and Pedagogical Content Knowledge
- Educator possesses strong technological and pedagogical content knowledge: The educator is highly proficient in using technology for teaching and has a deep understanding of the subject matter and effective teaching strategies. (4 points)
- Educator possesses adequate technological and pedagogical content knowledge: The educator is comfortable using technology for teaching and has a good grasp of the subject matter and teaching strategies. (3 points)
- Educator possesses limited technological and pedagogical content knowledge: The educator has some experience using technology for teaching but may lack

a strong foundation in the subject matter or effective teaching strategies. (2 points)

- Educator lacks necessary technological and pedagogical content knowledge: The educator has minimal experience using technology for teaching and lacks sufficient knowledge of the subject matter or effective teaching strategies. (1 point)

7. Acceptability and Usability for Target Student Population
   - Highly acceptable and usable for target student population: The AI tool's user interface and features are age-appropriate, engaging, and accessible for the target student population, considering their developmental stage and any special needs. (4 points)
   - Mostly acceptable and usable for target student population: The AI tool's user interface and features are generally appropriate and accessible for the target student population, with minor limitations or challenges. (3 points)
   - Somewhat acceptable and usable for target student population: The AI tool's user interface and features may not be fully age-appropriate or accessible for the target student population, presenting some barriers to effective use. (2 points)
   - Not acceptable or usable for target student population: The AI tool's user interface and features are not age-appropriate or accessible for the target student population, significantly hindering its usability and effectiveness. (1 point)

8. Psychometric Properties (if applicable for assessment)
   - Strong psychometric properties, providing reliable and valid assessment: The AI tool has been rigorously tested and demonstrates excellent reliability and validity in assessing student performance, aligning with established standards in the field. (4 points)
   - Adequate psychometric properties, providing mostly reliable and valid assessment: The AI tool has undergone some testing and shows acceptable levels of reliability and validity in assessing student performance, meeting most industry standards. (3 points)
   - Limited psychometric properties, with some concerns about reliability and validity: The AI tool has limited testing data available, and there are some concerns regarding its reliability and validity in assessing student performance, partially meeting industry standards. (2 points)
   - Poor psychometric properties, not suitable for reliable and valid assessment: The AI tool lacks sufficient testing data or has shown poor reliability and

validity in assessing student performance, failing to meet industry standards. (1 point)

K-12 Educator Self-Assessment Rubric for Implementing AI-Supported Collaborative Problem- Solving in STEM Classrooms

This self-assessment rubric is designed to help K-12 educators evaluate their readiness to implement AI-supported collaborative problem-solving activities in their STEM classrooms. The rubric assesses eight key areas: AI Knowledge and Skills, Technology Integration, Collaborative Learning, Assessment and Feedback, AI Ethics, Student Engagement, Professional Development, and Equity and Inclusion.

For each area, the rubric provides four levels of proficiency: Novice, Developing, Proficient, and Advanced. Educators can use the provided examples and evidence to self-identify their current level of readiness in each area. The rubric also includes a scoring system that allows educators to determine their overall readiness level, ranging from Novice to Advanced.

By using this rubric, educators can gain insights into their strengths and areas for improvement when it comes to integrating AI-supported collaborative problem-solving activities in their STEM classrooms.

AI Knowledge and Skills

1. Novice: I have limited understanding of AI concepts and how they can be used in education. I have not yet explored AI in my teaching or lesson planning.
2. Developing: I have a basic understanding of AI concepts and how they can be used in education. I have worked with lesson plans on AI topics that someone else developed.
3. Proficient: I have a good understanding of AI concepts and can apply them effectively in my teaching. I have created my own lesson plans that integrate AI concepts and tools.
4. Advanced: I have an in-depth understanding of AI concepts and can innovatively apply them in my teaching. I have designed a comprehensive AI curriculum and shared my expertise with colleagues.

Technology Integration

1. Novice: I rarely use technology in my teaching practices. I primarily rely on traditional teaching methods and materials.
2. Developing: I sometimes use technology in my teaching practices. I incorporate basic digital tools and resources in my lessons.
3. Proficient: I regularly use technology, including AI tools, in my teaching practices. I have successfully implemented AI-powered educational software in my classroom.
4. Advanced: I seamlessly integrate technology, including AI tools, into my teaching practices. I have developed innovative ways to use AI to enhance student learning and engagement.

Collaborative Learning

1. Novice: I rarely engage in collaborative professional learning or encourage collaborative problem-solving among my students. I mostly work independently and have students' complete tasks individually.
2. Developing: I sometimes engage in collaborative professional learning and encourage collaborative problem-solving among my students. I have participated in a few group projects or discussions related to AI in education.
3. Proficient: I often engage in collaborative professional learning and foster collaborative problem-solving among my students using AI tools. I have led a professional learning community on integrating AI in STEM education.
4. Advanced: I actively lead collaborative professional learning and effectively foster collaborative problem-solving among my students using AI tools. I have organized a school-wide initiative on AI-supported collaborative learning.

Assessment and Feedback

1. Novice: I rarely design assessments or provide feedback based on AI-supported collaborative activities. I primarily use traditional assessment methods.
2. Developing: I sometimes design assessments and provide feedback based on AI-supported collaborative activities. I have experimented with using AI tools to grade assignments and provide basic feedback.

3. Proficient: I regularly design effective assessments and provide constructive feedback based on AI-supported collaborative activities. I have created rubrics that evaluate students' performance in AI-based projects.

4. Advanced: I expertly design robust assessments and provide insightful feedback based on AI-supported collaborative activities. I have developed an adaptive assessment system that uses AI to personalize feedback for each student.

AI Ethics

1. Novice: I have limited understanding of AI ethics and responsible AI use. I have not yet considered the ethical implications of AI in education.

2. Developing: I have a basic understanding of AI ethics and responsible AI use. I have discussed AI ethics with my students but have not yet integrated it into my lesson plans.

3. Proficient: I have a good understanding of AI ethics and guide my students on responsible AI use. I have included discussions and activities on AI ethics in my lesson plans.

4. Advanced: I have an in-depth understanding of AI ethics and actively promote responsible AI use among my students. I have developed a comprehensive curriculum on AI ethics and responsible use.

Student Engagement

1. Novice: I rarely create engaging learning experiences or promote social and emotional learning through AI-supported collaborative activities. I primarily use traditional instructional methods.

2. Developing: I sometimes create engaging learning experiences and promote social and emotional learning through AI-supported collaborative activities. I have used AI-powered educational games in my classroom.

3. Proficient: I often create engaging learning experiences and effectively promote social and emotional learning through AI-supported collaborative activities. I have designed interactive AI-based simulations that foster teamwork and problem-solving skills.

4. Advanced: I consistently create highly engaging learning experiences and successfully promote social and emotional learning through AI-supported collaborative activities. I have developed an AI-supported learning environment that adapts to individual student needs and encourages collaboration.

Professional Development

1. Novice: I rarely participate in professional development programs related to AI integration and STEM education. I have not yet sought out opportunities to learn about AI in education.
2. Developing: I sometimes participate in professional development programs related to AI integration and STEM education. I have attended a few workshops or webinars on AI in education.
3. Proficient: I regularly participate in professional development programs to enhance my skills in AI integration and STEM education. I have completed a certification course on AI in education and regularly attend conferences on the topic.
4. Advanced: I actively seek out and lead professional development programs to enhance my skills in AI integration and STEM education. I have conducted workshops and presented at conferences on AI in STEM education.

Equity and Inclusion

1. Novice: I rarely consider how to make AI-supported activities inclusive for diverse student needs. I have not yet considered the potential biases in AI systems or their impact on different student populations.
2. Developing: I sometimes consider how to make AI-supported activities inclusive for diverse student needs. I have discussed the importance of diversity and inclusion in AI with my students but have not yet implemented specific strategies.
3. Proficient: I consistently ensure that AI-supported activities are inclusive and cater to diverse student needs. I have created differentiated AI-based learning activities that accommodate various learning styles and abilities.
4. Advanced: I actively promote and model best practices for ensuring AI-supported activities are inclusive and cater to diverse student needs. I have developed an equity-focused AI curriculum and collaborate with colleagues to ensure inclusive AI integration across the school.

Scoring and Results:

- 8-12 points: Novice readiness for implementing AI-supported collaborative problem-solving activities in STEM classrooms. Significant room for improvement and development.

- 13-20 points: Developing readiness for implementing AI-supported collaborative problem-solving activities in STEM classrooms. Some areas of strength, but still requires further development.
- 21-28 points: Proficient readiness for implementing AI-supported collaborative problem-solving activities in STEM classrooms. Well-prepared in most areas, with some room for enhancement.
- 29-32 points: Advanced readiness for implementing AI-supported collaborative problem-solving activities in STEM classrooms. Highly skilled and prepared to effectively integrate AI-supported collaborative learning in STEM education.

9

UNLEASHING CREATIVITY AND CRITICAL THINKING IN ARTS AND HUMANITIES EDUCATION

T he immersive Virtual Reality (VR) project "Becoming Homeless" allows users to experience life without a home, demonstrating AI's potential as a powerful tool for fostering social empathy and understanding complex societal issues. This project reflects a growing trend of AI intersecting with art, changing how the public engages with it. Such experiences can spark meaningful discussions and inspire students to think critically about social issues.

As we examine AI's impact on the arts, from VR and AR to algorithmic poetry, we are encouraged to consider new forms of creative expression in our digital future. Ms. Johnson, an art teacher, sees the potential for AI to revolutionize her classroom. She envisions her students collaborating with AI to create unique, thought-provoking pieces that push the boundaries of traditional art forms.

This chapter investigates how AI is not merely enhancing creativity but actively collaborating with it. It explores how this partnership can be leveraged to transform arts and humanities education, providing students with the tools and skills they need to thrive in an increasingly AI-driven world.

BALANCING AI-ASSISTED CREATIVITY AND HUMAN AUTHORSHIP

AI tools are changing how students learn, giving them new ways to explore and express their creativity. These tools can be especially useful for less experienced students, offering them different methods to share their ideas and develop their creative skills. However, the writing style of AI-generated content is not yet as good as human-authored work. Studies have found that AI-generated texts were seen as less well-written, inspiring, fascinating, interesting, and aesthetic compared to human-written and original text continuations. This highlights the need for guidelines on using AI as an inspirational aid rather than a replacement for human creativity.

The use of AI writing assistants blurs the lines between human and machine authorship, challenging traditional concepts of intellectual property rights. Many existing co-creative systems offer only one-way interaction, with humans interacting with AI but not vice versa. To foster engaging collaborative experiences, guidelines should encourage two-way communication between humans and AI. This shift in the understanding of creativity necessitates a reevaluation of the interplay between human ingenuity and AI-generated content.

Navigating this complex terrain requires guidelines that provide ethical guardrails for using generative AI as an inspirational tool rather than a replacement for personal expression. These guidelines could address issues such as prompting AI models, respecting intellectual property, and positioning AI as a supplemental creativity tool that augments human creativity. Collecting unintentional user data could enhance human-AI collaboration and user experience, potentially enabling AI to mimic human Theory of Mind abilities. Guidelines should specify the types of user data AI systems should gather to facilitate more natural collaboration. Moreover, when AI can explain its decision-making process and contributions, the system becomes more transparent and comprehensible, fostering trust and effective human-AI collaboration.

Educators should prioritize nurturing students' self-expression and original thinking while leveraging AI as a force multiplier for creativity. By adopting a proactive, principles-based approach, educators can create learning environments that encourage collaboration between students and AI systems, leading to the generation of novel ideas that exceed the original intentions of both parties. Drawing from research on effective human creative collaboration, guidelines for AI as an inspirational aid should be informed by factors that contribute to successful human collaboration.

The ultimate goal is to achieve a balance between AI-assisted creativity and human authorship, allowing students to benefit from the innovative possibilities offered by AI tools while preserving their unique voices and creative agency. By establishing clear guidelines and nurturing a collaborative learning environment, educators like Ms. Johnson can empower students to explore the potential of AI for creative expression and self-discovery. Through the thoughtful integration of AI tools and adherence to ethical principles, Ms.

Johnson envisions a classroom where students can push the boundaries of their creativity, cultivate their unique voices, and collaborate with AI systems to generate novel ideas that surpass what either could achieve independently.

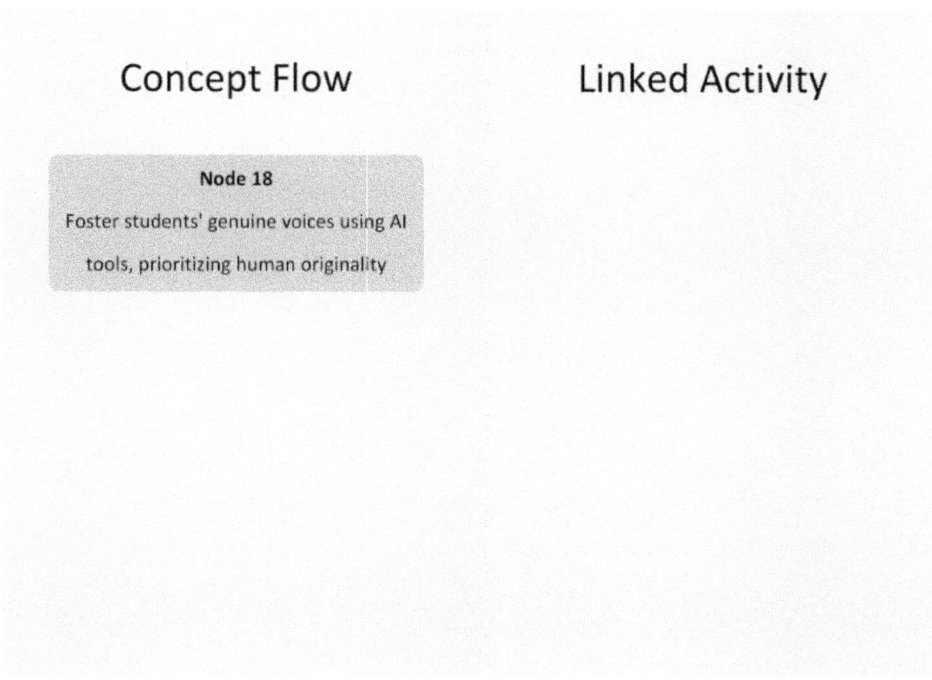

Callout Box: Unleashing Creative Potential: Insights for Educators

1. **Empowering Novice Creators:** AI tools offer less experienced students' novel ways to express their ideas and develop creative skills, although the stylistic quality may not yet match human-authored work.
2. **Redefining Authorship:** The use of AI writing assistants challenges traditional notions of intellectual property rights as the lines between human and machine authorship become increasingly blurred.
3. **Nurturing Authentic Voices:** Educators should prioritize fostering students' self-expression and original thinking while harnessing the power of AI to amplify their creativity.

NAVIGATING INTELLECTUAL PROPERTY RIGHTS IN THE AGE OF AI

Ms. Patel, a high school English teacher, stands in front of her class, a glimmer of excitement in her eyes. She introduces an AI-powered writing tool to her students, emphasizing that the tool is meant to inspire and support their writing process, not replace their unique voices. As she guides her students in using the AI tool to generate ideas and prompts, Ms. Patel encourages them to critically evaluate and build upon the AI-generated content to create original, thoughtful pieces that reflect their individual perspectives and experiences.

This real-world example highlights the exciting opportunities that integrating AI tools in educational settings presents for creative expression and engagement among students. AI tools open up new avenues for students to explore their creativity, potentially revolution-izing traditional educational practices. These tools provide unique ways for students, especially those with less experience, to articulate their ideas and enhance their creative skills.

However, as Ms. Patel's students begin to experiment with the AI writing tool, questions about the concept of authorship naturally arise. When a student uses an AI writing assistant to generate content, the lines between human and machine authorship become blurred, challenging conventional notions of intellectual property rights. This shift in the understanding of creativity necessitates a re-examination of the interplay between human creativity and AI-generated content.

Recognizing the need for clear guidelines and a collaborative learning environment, Ms. Patel develops a set of principles for her students to follow when using the AI writing tool. She emphasizes the importance of using the tool as a source of inspiration and support rather than relying on it to generate complete pieces of writing.

Ms. Patel also stresses the ethical considerations surrounding the use of AI-generated content, particularly in terms of intellectual property rights. She explains that while the AI tool can provide helpful prompts and ideas, the students must critically evaluate and build upon this content to create original works that reflect their own unique perspectives and experiences. By doing so, they can ensure that their writing remains authentically their own, even if AI-generated suggestions inspired it.

To foster a collaborative learning environment, Ms. Patel encourages her students to share their experiences using the AI writing tool and discuss the challenges and opportunities it presents. Through these discussions, the students gain a deeper understanding of the complex relationship between human creativity and artificial intelligence and learn how to navigate this new landscape in an ethical and responsible manner.

As the class progresses, Ms. Patel witnesses her students' growth and the emergence of unique, creative pieces that showcase their individual talents and experiences. The AI writing tool becomes a catalyst for discussion and collaboration, with students sharing their insights and ideas, learning from one another, and pushing the boundaries of their creativity. By emphasizing the importance of critical thinking, originality, and ethical considerations, Ms. Patel helps her students develop the skills and mindset necessary to thrive in a world where AI plays an increasingly significant role in creative processes.

Callout Box: Navigating Intellectual Property Rights in the Age of AI

1. **Establishing Ethical Guidelines:** As AI-generated art presents new opportunities and challenges, future research should focus on developing ethical guidelines, exploring collaborative practices, and examining the social and political implications of AI-based art.
2. **Designing Effective Co-Creative Systems:** Interaction design is critical for creating effective human-AI co-creative systems. Frameworks like COFI can guide designers in exploring the design space of interaction and analyzing existing systems to identify trends and gaps.
3. **Addressing Ethical Concerns:** The use of AI in art raises ethical issues around intellectual property rights, privacy, and bias. The ownership of AI-generated works and the rights of original creators and programmers remain topics of ongoing debate.

PROMOTING ETHICAL AI LITERACY IN ARTS AND HUMANITIES CLASSROOMS

As AI-powered tools like Google's DeepDream, OpenAI's MuseNet, DeepArt, and Artbreeder gain traction in classrooms, educators must address the complex issues surrounding intellectual property rights head-on. Mr. Nguyen, a middle school art teacher, recognizes this need and develops a lesson plan exploring the ethical implications of using AI-generated art when introducing an AI-powered image-generation tool to his students.

These AI tools offer exciting opportunities for students to explore their creativity and engage in collaborative learning, regardless of their skill level. DeepArt and Artbreeder, for instance, provide user-friendly interfaces that allow students to experiment with artistic styles and create unique works of art. Yet, the use of AI-generated content raises questions about originality, authorship, and the ethical implications of using such tools.

Mr. Nguyen understands the importance of integrating AI ethics education alongside technical skills. He designs a transdisciplinary curriculum that fosters critical sociotechnical competencies and broadens participation through intentional equity-centered instructional design. By engaging students in collaborative activities that wrestle with issues of bias, discrimination, surveillance, and autonomy, Mr. Nguyen aligns his approach with the recommendations to teach students to recognize the ethical challenges and implications of algorithm use beyond just technical aspects.

Updated frameworks that balance innovation and protection are necessary to navigate this complex landscape. Mr. Nguyen takes a proactive approach, integrating intellectual property education into his curriculum and emphasizing the importance of respecting copyrights, trademarks, and patents. He focuses on developing his students' critical thinking skills when using AI in art, encouraging them to reflect on the ethical implications of AI-generated works.

Mr. Nguyen investigates how an AI art curriculum can engage younger students with key AI concepts in an age-appropriate way. He recognizes the feasibility and value of developing pretertiary AI curricula. He incorporates elements of the draft framework to ensure a comprehensive ethics of AI in education that synthesizes the ethics of education, learning sciences, data, and algorithms.

By creating a supportive learning environment that fosters discussions on intellectual property rights and encourages students to cite and reference sources properly, Mr. Nguyen raises awareness about the significance of respecting the work of others. He utilizes visual arts and creative projects to engage his students in discussions about intellectual property rights, allowing them to explore the intersection of art, technology, and ethics. Additionally, implementing practical tools like watermarking systems serves as a hands-on approach to teaching his students about respecting intellectual property rights.

Mr. Nguyen's approach exemplifies the best practices for educators promoting ethical AI literacy in arts and humanities classrooms. These practices include:

1. Integrating AI ethics education alongside technical skills
2. Designing transdisciplinary curricula that foster critical sociotechnical competencies
3. Engaging students in collaborative activities that wrestle with issues of bias, discrimination, surveillance, and autonomy
4. Developing students' critical thinking skills when using AI in art
5. Utilizing visual arts and creative projects to explore the intersection of art, technology, and ethics
6. Implementing practical tools like watermarking systems to teach students about respecting intellectual property rights

Recognizing the need for more professional development for K-12 teachers, Mr. Nguyen engages in reflective teaching practices. He participates in a community of practice to share experiences and enhance his instructional practices on AI ethics. He models ethical behavior and provides students with direct experience in grappling with ethical issues.

Educators like Mr. Nguyen can effectively promote ethical AI literacy in arts and humanities classrooms by adopting these best practices and embracing a transdisciplinary approach to AI education.

Callout Box: Promoting Ethical AI Literacy in Arts and Humanities Classrooms

1. **Transdisciplinary Approach:** Foster critical sociotechnical competencies and broaden participation by integrating AI ethics and technical learning through collaborative activities addressing bias, discrimination, surveillance, and autonomy.
2. **Age-Appropriate Curriculum:** Develop pretertiary AI curricula that engage younger students with key AI concepts, incorporating ethics of education and learning sciences alongside data and algorithms.
3. **Teacher Empowerment:** Enhance K-12 teachers' instructional practices on AI ethics through reflective teaching, communities of practice, and modeling ethical behavior while providing students with hands-on ethical problem-solving experiences.

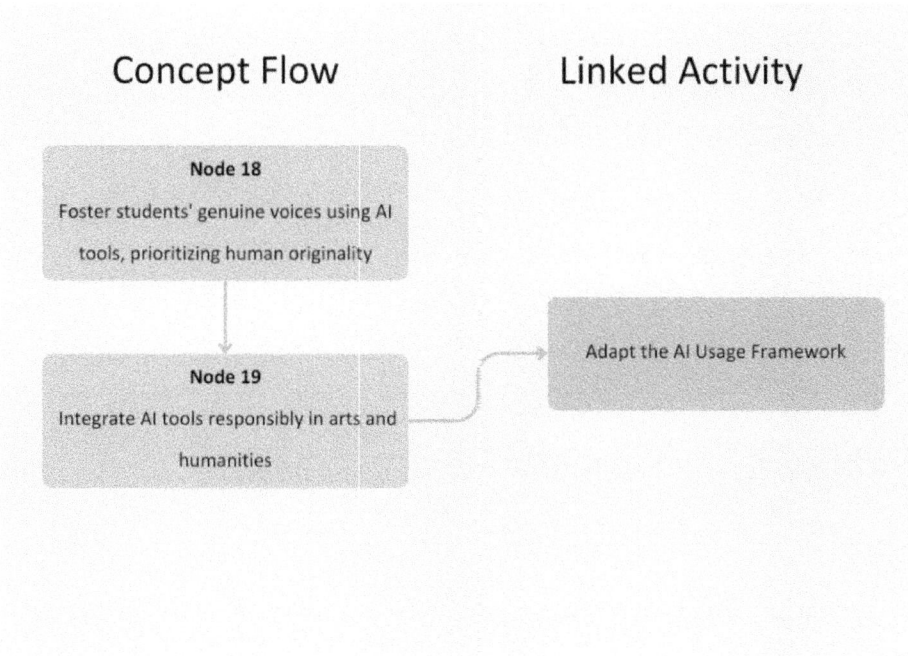

ADAPTING THE AI USAGE FRAMEWORK FOR ARTS AND HUMANITIES

Ms. Patel, a high school art teacher, was excited to introduce an AI-powered image-generation tool to her students. As she began to plan how to incorporate this new technology into her lessons, she recalled the AI usage framework from Chapter 2, which provided a roadmap for educators to integrate AI into their teaching responsibly.

Adapting this framework to her art classes, Ms. Patel considered the unique challenges AI presented in the arts and humanities. She knew that human interpretation and the analysis of human-created works were central to these subjects, and she wanted to ensure that her students would use the AI tool responsibly.

At the framework's core was the principle of "No Raw AI Output." Ms. Patel understood that she would need to carefully review and customize any AI-generated content to ensure accuracy and alignment with her learning objectives. She made a note to discuss this principle with her students, emphasizing the importance of their own creative input and critical thinking skills.

As she delved deeper into the framework, Ms. Patel considered how she could apply the different levels of AI integration in her art classes. She knew that Level 0 (Red) prohibited AI use in high-stakes assessments, such as generating project ideas for summative evaluations. To maintain integrity, these assessments would need to remain entirely human-authored.

Moving on to Level 1 (Yellow), Ms. Patel saw an opportunity to use AI for research and ideation. She could use the AI tool to gather resources for lesson planning, such as generating a list of relevant artistic styles or techniques for a unit on a specific art movement. However, she reminded herself that directly incorporating AI-generated lesson plans without significant editing would be unacceptable.

At Level 2 (Green), Ms. Patel envisioned using AI as a support tool. She could generate drafts of worksheets or visual aids, which she would then heavily revise and personalize for her students. For example, she could use the AI tool to create a rough sketch of a still-life composition and then refine it for a drawing lesson. She made a note to disclose any AI assistance to her students clearly.

Finally, at Level 3 (Green), Ms. Patel considered how AI could act as a teaching assistant. She could explore vetted AI-powered tools that provide interactive support during instruction, such as a platform that allows students to engage in virtual conversations with famous artists. However, she knew that relying on AI to deliver core instruction or using unapproved platforms autonomously would be unacceptable.

As she introduced the AI image generation tool to her students, Ms. Patel led a discussion on the responsible use of AI in art. She encouraged her students to respect intellectual property rights and to use the tool to enhance their own creativity, not replace it. By setting clear guidelines and fostering critical thinking, Ms. Patel empowered her students to harness the benefits of AI while nurturing their unique artistic voices.

Through her experience adapting the AI usage framework for her art classes, Ms. Patel realized that the key to successfully integrating AI in arts and humanities education lay in carefully considering the unique ethical challenges these subjects pose. By adhering to the "No Raw AI Output" principle and thoughtfully applying the framework across different levels of AI integration, she could leverage AI to enhance teaching and learning while upholding the core values of human-led instruction, academic integrity, and equitable access to high-quality education.

Callout Box: Adapting the AI Usage Framework for Arts and Humanities

1. **Core Principle:** Adhere to the "No Raw AI Output" principle, carefully reviewing and customizing AI-generated content to ensure accuracy and alignment with learning objectives.
2. **Levels of AI Integration:** Apply the AI usage framework across different levels:
   - Level 0 (Red): Prohibit AI use in high-stakes assessments to maintain integrity.
   - Level 1 (Yellow): Use AI for research and ideation, such as gathering resources for lesson planning.
   - Level 2 (Green): Employ AI as a support tool, generating drafts of worksheets or visual aids that are heavily revised and personalized.
   - Level 3 (Green): Explore vetted AI-powered tools that provide interactive support during instruction while avoiding reliance on AI for autonomous core instruction delivery.
3. **Ethical Considerations:** Lead discussions on the responsible use of AI in art, encouraging students to respect intellectual property rights and use AI tools to enhance rather than replace their creativity.

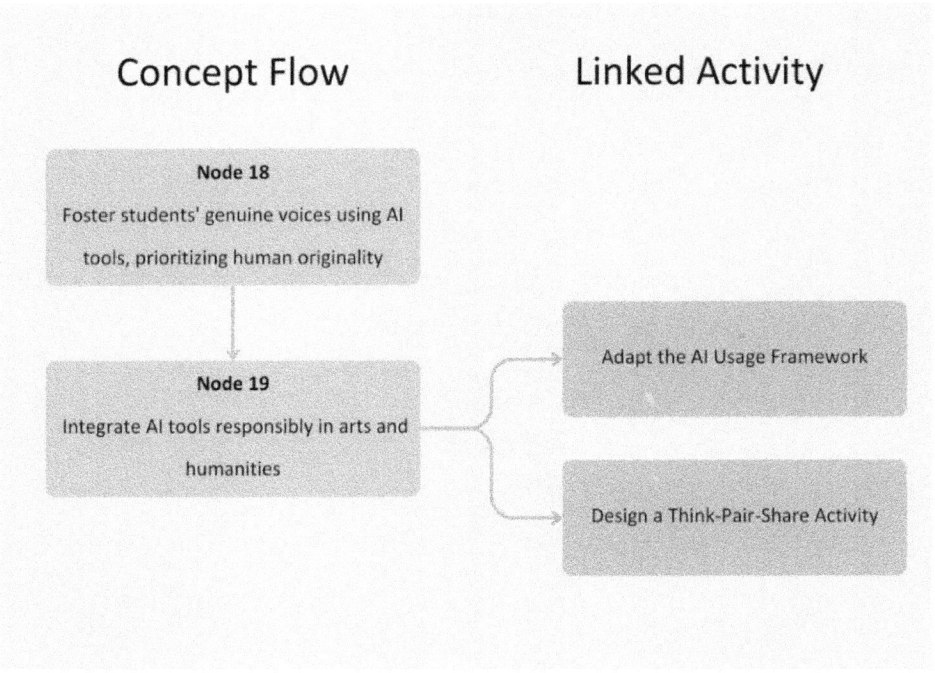

RESULT EXAMPLE OF THINK-PAIR-SHARE ACTIVITY: EXPLORING THE INFLUENCE OF ARTIFICIAL INTELLIGENCE ON HISTORICAL INVESTIGATION

In the following section, we present a step-by-step guide for educators to create engaging AI-generated activities that promote critical thinking and discussion about AI's impact on various disciplines. To demonstrate the effectiveness of this process, we have applied these steps to develop a Think-Pair-Share activity that explores the influence of artificial intelligence on historical investigation. This example serves as a practical illustration of how AI can be leveraged to enhance student engagement and foster a deeper understanding of AI's role in shaping our world.

The output of the below steps is in the appendix as "Example: Think-Pair-Share Activity: Exploring AI's Influence on Historical Investigation."

Step 1: Defining the Activity and Objectives

Generic Guide: First, identify your learning goals. What do you want students to learn from the activity? Pick an activity format that matches these goals, using ChatGPT for ideas if needed. This makes sure the activity's design supports your educational objectives.

Our Application: We wanted students to critically evaluate AI's impact on historical research and recognize biases in AI narratives. Talking with ChatGPT helped us choose the Think-Pair-Share model, which encourages thoughtful reflection, discussion, and teamwork.

Example Prompt: "Looking for an engaging activity for high schoolers to think critically about AI's role in history. Main goals: critical evaluation of AI in historical research and understanding biases. Ideas for formats and addressing common AI misconceptions?"

Step 2: Drafting the Initial Prompt for AI

Generic Guide: Write a clear, concise AI prompt that describes the planned activity and its educational goals. Mention the student age group, key themes to explore, and any areas that might be confusing.

Our Application: We wrote a prompt to generate questions about AI's use in historical research for high school students, pointing out areas that needed more explanation.

Example Prompt: "I'm preparing a Think-Pair-Share activity on AI in historical research. Goal: encourage critical discussion on AI's impact and biases. Need thought-provoking questions and tips on addressing AI misconceptions. Ideas?"

Step 3: AI Engagement and Iteration

Generic Guide: Begin with an initial AI request that fits your activity's structure. Check how well the AI's response matches your educational goals, adjusting your approach as needed to improve the content.

Our Application: After reviewing the AI's first set of questions, we wanted more specific questions about ethics and biases in AI history. By going back and forth with ChatGPT, we came up with better, more engaging questions and discussion strategies.

Step 4: Finalizing the Activity

Generic Guide: Shape the AI's suggestions into a detailed activity plan that encourages deep thinking and participation. Create additional resources to support full student engagement.

Our Application: We adapted the format of the previous activity to our current focus on AI in history, adding ChatGPT's ideas on debating AI's benefits and risks and preparing extra materials for an in-depth discussion.

Teacher's Innovation: "Adapting last term's environmental debate for our AI history topic. Need to shift the debate to AI's historical roles and biases. Also, making supplementary materials for a well-rounded debate. Ideas?"

Step 5: Implementing and Reflecting on the Activity

Generic Guide: Run the activity with clear instructions and support. After the activity, assess its success and ask for student feedback to keep improving.

Our Application: We led the Think-Pair-Share activity, making sure each phase was understood. Afterward, we reflected on how well it worked and gathered student feedback for future improvements.

Example AI Final Prompt: "Design a Think-Pair-Share on AI's impact in historical research for 10-11th graders. Focus on critical analysis of AI's research role, benefits, drawbacks, and biases. Need engaging questions and strategies for leading the discussion."

EXAMPLE: THINK-PAIR- SHARE ACTIVITY: EXPLORING AI'S INFLUENCE ON HISTORICAL INVESTIGATION

Step 1: Think

Take a moment to think about artificial intelligence and how it affects historical research.

Consider these questions:

- How do you think AI can be used in historical research?
- What are some possible benefits of using AI in this field?
- Are there any potential drawbacks or worries about depending on historical accounts made by AI?
- How can biases show up in historical accounts made by AI?

Step 2: Pair

Find a partner and share your thoughts and ideas on the questions from Step 1.

Listen carefully to your partner's responses and ask follow-up questions if needed.

Have a meaningful discussion, talking about the different perspectives and insights you each gained from the other's responses.

Step 3: Share

With your partner, share your main takeaways from the discussion with the rest of the class.

Volunteer to share your thoughts on any of the questions from Step 1.

Encourage others to participate by asking additional questions or looking for different viewpoints.

Take notes on the various perspectives shared during the class discussion.

This activity will allow students to evaluate how artificial intelligence influences historical research critically. By having a thoughtful discussion and sharing ideas, students will be able to identify potential biases in historical accounts generated by AI

10

UNLOCKING THE POWER OF PROMPT ENGINEERING

A PRIMER FOR K-12 EDUCATORS

Creating effective prompts is key to unlocking AI-assisted learning in K-12 education. It is also why we end our book here, as it represents the culmination of the concepts and strategies we've explored. A well-designed prompt steers the interaction between students and AI systems, ensuring meaningful outcomes, just as a lesson plan guides students.

The essential elements of effective prompts include:

- Clarity and specificity: Clear prompts help students understand expectations and enable AI models to generate relevant responses.
- Contextual relevance: Prompts should align with learning objectives and subject matter, making AI interactions an integral part of the educational experience.
- Personalization: Adaptable prompts cater to each student's unique learning needs and preferences, unlocking their potential.
- Collaboration and teamwork: Prompts encouraging collaboration fostered a sense of community and shared discovery.
- Ethical considerations: Prompts should promote responsible AI use and raise awareness about ethical implications.
- Reflection: Reflective prompts allow students to evaluate their thinking processes and knowledge, promoting growth and development.

Callout Box: Real-World Examples: Applying Essential Elements in the Classroom

1. Clarity and specificity: Example: "Write a 150-word summary of the main character's journey in the novel." Benefit: Students focus on key aspects, developing concise writing skills.
2. Contextual relevance: Example: "Using AI, generate a dialogue between two historical figures discussing a pivotal event." Benefit: Students engage with the subject matter, deepening their understanding of historical context.
3. Personalization: Example: "Create a poem using AI, incorporating themes that resonate with your personal experiences." Benefit: Students connect with the material on a personal level, increasing motivation and engagement.
4. Collaboration and teamwork: Example: "As a group, use AI to generate a mind map of the ecosystem, then discuss and refine it together." Benefit: Students learn to work collaboratively, building communication and problem-solving skills.
5. Ethical considerations: Example: "Analyze an AI-generated article for potential biases and discuss the implications." Benefit: Students develop critical thinking skills and awareness of ethical issues in AI.
6. Reflection: Example: "Use AI to create a visual representation of your learning journey, then reflect on your growth." Benefit: Students develop metacognitive skills, taking ownership of their learning process.

TAILORING PROMPTS FOR DIFFERENT LEARNING OBJECTIVES

Effective, prompt engineering in K-12 education requires tailoring prompts to specific learning objectives.

Brainstorming

When designing brainstorming prompts, use open-ended questions to encourage creativity and idea generation. These prompts should inspire students to explore multiple perspectives and question assumptions, facilitating expansive thinking. By encouraging thinking outside the box, educators can foster a learning environment that values innovation and original thought.

Tip: Design open-ended prompts that encourage students to explore multiple perspectives and generate creative ideas. Start prompts with phrases like "How might we…" or "What if…" to facilitate expansive thinking.

Feedback

When giving feedback, prompts should encourage students to reflect on their work critically and seek feedback. These prompts should guide students in articulating their thoughts clearly, enabling them to provide and receive specific, actionable feedback. By using feedback loops, educators can help students develop the skills to evaluate their work and the work of others effectively.

Tip: Create prompts that guide students in providing constructive feedback to their peers. Encourage students to use these prompts to articulate their thoughts clearly and offer specific, actionable suggestions. This approach, informed by research, helps students develop effective feedback skills.

Critical Thinking

Educators should design prompts that require analytical thinking, evaluation, and synthesis of information to encourage critical thinking. These prompts should challenge students to scrutinize AI-generated content and consider multiple viewpoints, enhancing their ability to engage in thoughtful analysis. By exposing students to diverse perspectives and encouraging them to question assumptions, educators can help develop the critical thinking skills essential for 21st-century success.

Tip: Incorporate AI-generated content into your lessons and develop prompts that challenge students to analyze the material critically. Design questions that encourage students to question assumptions, consider alternative viewpoints, and engage in thoughtful analysis..

Personalization and Context

Tailoring prompts for learning objectives requires personalization and context. Prompts should align with individual learning objectives, students' interests, and classroom contexts to ensure relevance and engagement, increasing educational value. Educators should ensure that prompts are contextually relevant and tailored to individual needs. Educators can increase engagement and motivation by creating prompts that resonate with students, leading to better learning outcomes.

Tip: Conduct surveys or questionnaires to gather information about your students' interests, learning preferences, and background knowledge. Use this data, along with provided insights, to create prompts that are tailored to your student's needs and relevant to their context, increasing engagement and motivation.

Integration of Feedback

Effective, prompt engineering requires the integration of feedback. Educators should include mechanisms for iterative feedback from AI to help students refine their understanding and responses. This interaction helps students learn and improve their critical thinking and problem-solving skills. By allowing students to engage with AI systems, educators can create a more dynamic learning experience.

Tip: Implement an iterative feedback system where students submit their responses to AI-generated prompts and receive automated feedback. Design prompts that encourage students to refine their understanding and responses based on this feedback, creating a dynamic learning experience that enhances critical thinking and problem-solving skills.

Collaborative Learning

Prompts should promote collaborative learning by encouraging students to formulate responses and interpret AI outputs. These prompts foster teamwork and collective problem-solving skills. By designing prompts that require collaboration and knowledge, educators can create an environment of cooperation and mutual support.

Tip: Develop prompts that require students to collaborate in analyzing AI-generated content, solving problems, or generating new ideas. Design tasks that assign specific roles and responsibilities within student groups to ensure active participation and knowledge sharing, fostering a cooperative and supportive learning environment.

ENSURING AGE-APPROPRIATENESS, BIAS-FREE CONTENT, AND ETHICAL ALIGNMENT

When designing prompts for K-12 students, it's crucial to consider age-appropriateness, bias-free content, and ethical alignment. Prompts should match different age groups' cognitive and emotional maturity by using appropriate language and themes and ensuring engaging and accessible content to maintain attention and motivation. By tailoring prompts to each age group's needs and abilities, educators create a challenging and supportive learning environment.

Mitigating bias is essential in prompt engineering. research has emphasized eliminating potential biases related to gender, race, ethnicity, or other personal characteristics. This involves reviewing and removing biased language or content and incorporating diverse perspectives for inclusivity and cultural sensitivity. By creating bias-free prompts, educators can ensure equal opportunities for all students to engage and express their ideas.

Adhering to ethical guidelines is critical when designing prompts for K-12 students. Good practices include obtaining informed consent for using student responses, respecting privacy, and ensuring voluntary participation. Ethical considerations also include designing prompts that don't harm the student's emotional or psychological state and promoting a respectful and safe learning environment. Prioritizing ethical considerations helps create an engaging and responsible learning environment.

Promoting critical thinking is a key aspect of prompt engineering. Prompts should encourage students to engage in thinking that involves analyzing information critically and exploring multiple perspectives. Encouraging students to question assumptions and evaluate source credibility can cultivate a reflective approach to learning, which is crucial. Educators can help students develop skills to navigate complex issues and make informed decisions by designing prompts that promote critical thinking.

Prompts should encourage students to reflect on the ethical implications of their knowledge. This type of reflection can be facilitated by asking students to consider the broader impacts of their learning and its application to real-world scenarios, enhancing their understanding and ethical awareness. Incorporating reflective elements into prompts can help students develop a deeper understanding of the material and its relevance to their lives and the world.

Ensuring age-appropriate, bias-free, and ethical prompts is essential for K-12 students. Educators can create an engaging, inclusive, and responsible learning environment by tailoring prompts to the specific needs and abilities of each age group, mitigating bias, adhering to ethical guidelines, promoting critical thinking, and facilitating reflection.

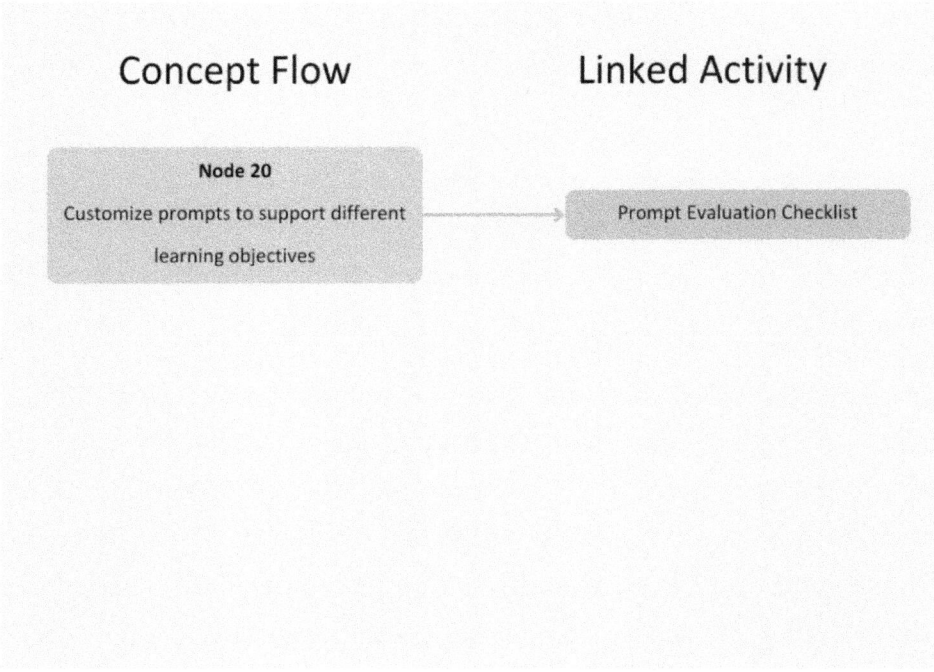

Prompt Evaluation Checklist: Ensuring Age-Appropriateness, Bias-Free Content, and Ethical Alignment

Language and Complexity:

- Is the language used in the prompt appropriate for the student's age group?
- Is the prompt's complexity and length appropriate for the student's attention span and reading level?

Cognitive and Emotional Development:

- Are the concepts and themes addressed in the prompt suitable for the student's cognitive and emotional development?
- Does the prompt engage students at their current level of understanding while still challenging them to grow?

Diversity and Inclusion:

- Does the prompt avoid language or content that perpetuates stereotypes or prejudices based on race, gender, ethnicity, religion, or other personal characteristics?

- Are diverse perspectives and experiences represented in the prompt's content and examples?
- Does the prompt encourage students to consider multiple viewpoints and challenge their own biases?

Ethical Guidelines and Principles:

- Does the prompt adhere to the ethical guidelines and principles established by the school or educational institution?
- Is the use of AI in the prompt transparent, and are students informed about how their data will be used and protected?
- Does the prompt avoid encouraging or endorsing unethical behavior, such as plagiarism, cheating, or deception?

Risk Assessment and Mitigation:

- Are the potential risks and benefits of using AI in the learning activity addressed and discussed with students?
- Does the prompt foster a safe and inclusive learning environment that respects students' privacy and individual differences?

Critical Thinking Skills:

- Does the prompt encourage students to analyze information critically and question assumptions?
- Are students prompted to evaluate the credibility and reliability of sources, including AI-generated content?
- Does the prompt challenge students to consider the implications and consequences of their ideas and decisions?

Reflective Practice:

- Does the prompt encourage students to reflect on their learning process and personal growth?
- Are students prompted to consider the real-world applications and ethical implications of their newfound knowledge?

- Does the prompt create opportunities for students to provide feedback on their learning experience and suggest improvements?

PROMPT ENGINEERING TECHNIQUES FOR K-12 EDUCATORS

Prompt engineering techniques are essential tools for K-12 educators looking to use AI-assisted learning. Teachers can use prompting strategies to guide students through complex concepts, promote critical thinking, and foster understanding.

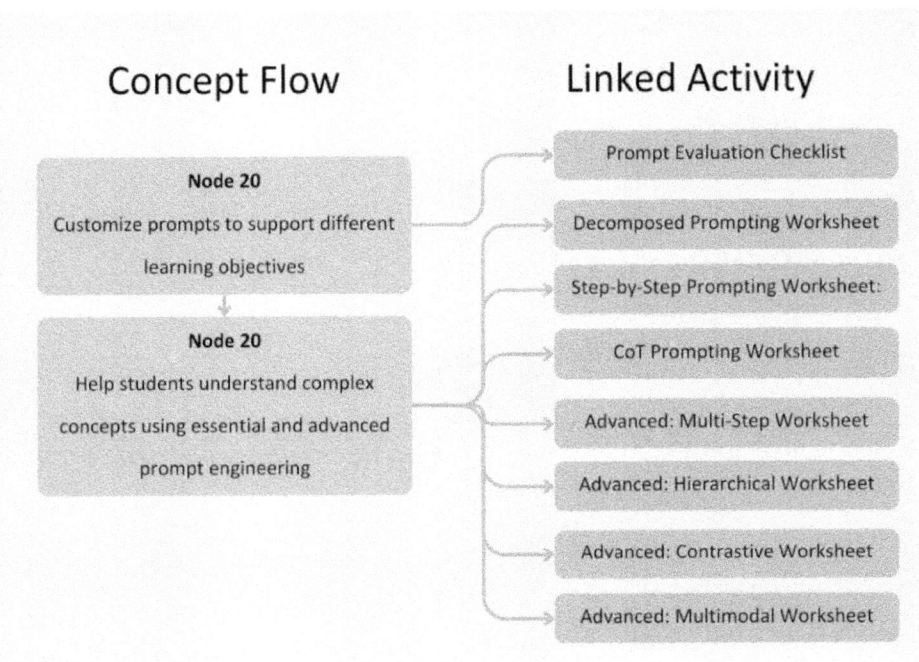

Essential Techniques

Decomposed prompting

Decomposed prompting is a technique that involves breaking down complex questions into more manageable sub-questions or steps. This approach clarifies the reasoning process and improves student responses by providing a structured and sequential thought process.

K-12 Application: Educators can use decomposed prompting to help students tackle complex problems in various subjects, such as math, science, or literature. By guiding students through a series of sub-questions or steps, teachers can facilitate deeper understanding and retention. This technique provides support that gradually fades as students become more proficient, with each step building upon the previous one, ultimately enhancing critical thinking and problem-solving skills.

Step-by-Step Prompting

Step-by-step prompting is a technique that guides students through a logical sequence of steps to solve problems or understand concepts. This method is beneficial in structured learning environments where tasks are sequenced, promoting a progressive learning atmosphere ideal for skill development and understanding.

K-12 Application: Educators can use step-by-step prompting to help students solve multi-step math problems in a 4th-grade classroom. By providing a clear, sequential set of instructions, teachers can guide students through the problem-solving process, ensuring that they understand each step and can apply the necessary skills. This technique is particularly useful for complex problems that require multiple operations or concepts, as it helps students organize their thoughts and approach the problem systematically.

Chain of Thought Prompting

Chain of Thought Prompting is a technique that guides students to articulate their reasoning step-by-step before reaching a conclusion. This method is valuable for tasks requiring logical deduction or complex reasoning, making it beneficial for educational purposes where understanding the thinking process is crucial. Chain of Thought Prompting encourages students to follow a structured thinking process, incorporating intermediate steps and enhancing their reasoning abilities.

K-12 Application: Educators can use Chain of Thought Prompting to help 4th-grade students solve multi-step math problems by encouraging them to explain their reasoning at each step. This technique allows teachers to assess students' understanding of the problem-solving process and identify areas where they may need additional support. Students can develop a deeper understanding of the concepts involved and improve their problem-solving skills by articulating their thought process.

Advanced Techniques

Multi-step prompting

Multi-step prompting involves interconnected steps that guide the learner through a complex learning trajectory. This technique enhances comprehension and critical thinking by breaking down information and requiring the logical linking of parts. Multi-step prompting fosters deeper engagement and is particularly effective in problem-solving or detailed analysis tasks.

K-12 Application: In a K-12 setting, multi-step prompting can be used to guide students through complex problem-solving tasks or in-depth analyses of literary works, historical events, or scientific phenomena. Educators can help students break down the problem or topic into more manageable parts by providing a series of interconnected prompts, encouraging them to think critically and make logical connections between each step.

For example, in a middle school English class, a teacher can use multi-step prompting to guide students through an analysis of a short story. The prompts may include identifying the main characters, describing the setting, analyzing the plot, interpreting the theme, and evaluating the author's writing style. By following these interconnected steps, students can develop a comprehensive understanding of the story and improve their analytical skills.

Hierarchical prompting

Hierarchical prompting organizes learning tasks into a structured hierarchy, supporting a natural learning progression. This method allows educators to tailor learning experiences to students' needs and capabilities, promoting systematic problem-solving and comprehensive subject understanding.

K-12 Application: Hierarchical prompting can be used in a K-12 setting to guide students through complex tasks or concepts in various subjects, such as science, math, or language arts. By organizing prompts into a structured hierarchy, educators can help students progress from foundational knowledge to more advanced understanding and application.

For example, in a high school biology class, a teacher can use hierarchical prompting to guide students through an exploration of the human digestive system. The prompts may start with basic anatomy and progress to more complex topics like digestive processes, nutrient absorption, and the relationship between diet and health. By following this hierarchical structure, students can develop a comprehensive understanding of the subject matter and improve their problem-solving skills.

Socratic prompting

Socratic prompting uses questions to stimulate critical thinking and reflection. Educators can enhance analytical skills, promote active participation, and foster understanding of complex concepts by challenging students to think deeply. This method is useful in discussions and debates, encouraging students to articulate their thoughts and consider multiple perspectives.

K-12 Application: In a K-12 setting, Socratic prompting can be used to facilitate meaningful discussions and encourage critical thinking across various subjects. By posing thought-provoking questions, educators can help students explore complex ideas, challenge assumptions, and develop a deeper understanding of the subject matter.

For example, in a high school English class, a teacher can use Socratic prompting to guide students through a discussion of a literary work. The teacher may ask questions that encourage students to analyze characters, interpret themes, and consider the author's purpose. By engaging in this type of dialogue, students can develop their critical thinking skills, improve their ability to articulate their thoughts, and gain a more comprehensive understanding of the text.

Specialized Techniques

Contrastive Prompting

Contrastive Prompting is a technique that uses contrasting elements or prompts to guide learning or reasoning. By presenting contrasting examples or prompts, this method highlights differences, similarities, or specific features, aiding in understanding and differentiating concepts.

Research has explored the effectiveness of Contrastive Prompting in various fields, demonstrating its potential to enhance learning outcomes and cognitive processes. In K-12 education, Contrastive Prompting can be valuable for educators using platforms like ChatGPT to facilitate student learning. By providing contrasting examples or prompts, educators can help students distinguish between different concepts, improve comprehension, and foster critical thinking skills. For example, contrastive video examples in teacher education have shown promise in promoting the acquisition of educational knowledge by future teachers.

Contrastive Prompting has also been studied in language learning. Studies have shown proposed automated trans-lingual definition generation through contrastive prompt learning to assist language learners. This approach aims to enhance language understanding and vocabulary acquisition among learners.

In online security, Contrastive Prompting has been explored to predict the structure of online discussions about security topics in online forums. This was done by using short and contrasting prompts to improve the prediction accuracy of discussion structures, demonstrating the applicability of this technique in complex tasks.

K-12 Application: In a K-12 setting, Contrasting Prompting can help students understand and differentiate between various concepts, such as literary devices, historical events, or scientific phenomena. By presenting contrasting examples or prompts, educators can guide students in analyzing the differences and similarities between the concepts, fostering critical thinking and deeper understanding.

For example, in an English literature class, a teacher can use Contrastive Prompting to help students distinguish between different literary devices, such as metaphors and similes. By providing contrasting examples of each device and asking students to analyze the differences, the teacher can facilitate a better understanding of the unique characteristics and purposes of each literary device.

Multimodal prompting

Multimodal prompting uses different modes of communication or stimuli to elicit responses. It can be an effective way to engage students and enhance learning in K-12 education.

Multimodal prompting in education often involves using visual aids in conjunction with verbal or written prompts. This approach makes learners' perspectives more visible, helping educators understand and address their needs more effectively. Incorporating images, videos, or interactive elements alongside text-based prompts can help students grasp complex concepts and stay engaged.

Multimodal prompting can be particularly effective for students with diverse learning styles or needs. By presenting information and eliciting responses through multiple channels, educators can cater to different preferences and abilities, ensuring a more inclusive learning environment.

When using multimodal prompting in education, it is crucial to consider the ethical implications since it highlights the need to reevaluate multimodal design practices to ensure that they are appropriate, accessible, and do not disadvantage any students.

When designing multimodal prompts for AI-assisted learning using platforms like ChatGPT 4 or Claude, which can process more than just text, educators should:

1. Choose modes that align with learning objectives and student needs.
2. Ensure accessibility and inclusivity for all learners.
3. Consider the ethical implications of the chosen modes.
4. Provide clear instructions for engaging with multimodal prompts.

5. Monitor student responses and adapt prompts as needed.

By using multimodal prompting effectively, educators can leverage AI to create engaging, effective, and inclusive learning experiences for their students.

K-12 Application: In a K-12 setting, multimodal prompting can be used to create engaging and interactive learning experiences across various subjects. By incorporating visual aids, audio clips, or interactive elements alongside text-based prompts, educators can cater to different learning styles and keep students motivated.

For example, in a science class, a teacher can use multimodal prompting to help students understand the water cycle. The teacher can provide an image of the water cycle, along with text-based prompts asking students to describe each stage. Students can then use an AI tool like ChatGPT 4 or Claude to generate their responses, incorporating the visual information and their own understanding. This approach allows students to engage with the content in multiple ways, deepening their understanding and retention of the material.

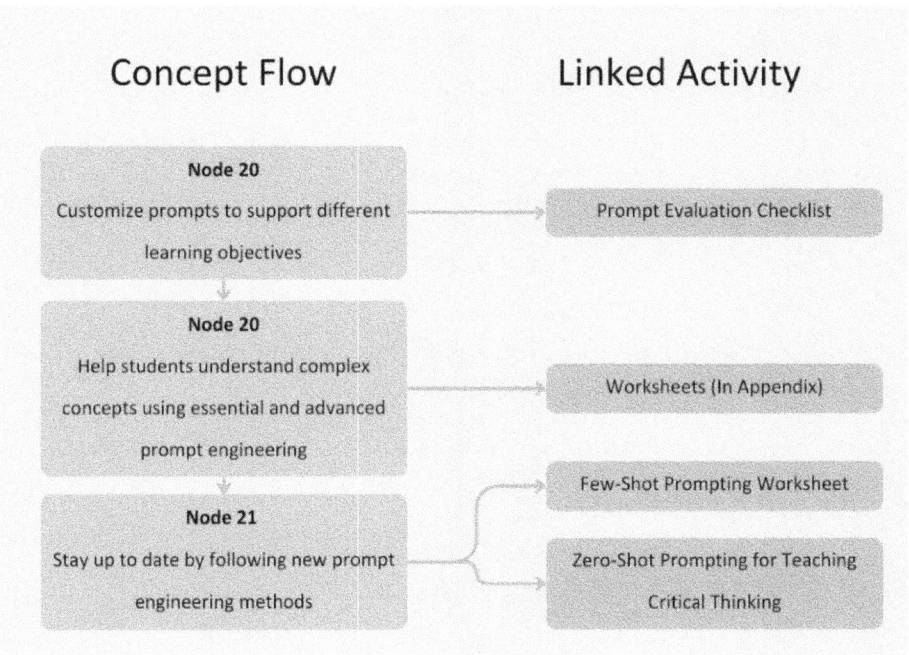

Emerging Techniques

Zero-Shot Prompting

Zero-shot prompting is a technique that enables AI models like ChatGPT to respond to queries based on general knowledge without requiring specific prior examples from the task domain. This allows the AI to respond rapidly to a wide range of questions, making it a valuable tool in educational settings.

K-12 Application: Zero-shot prompting can effectively teach critical thinking skills in a K-12 setting. Educators can encourage students to formulate complex questions, predict AI responses, and evaluate the accuracy and depth of the AI-generated answers. This process helps students develop skills in anticipating AI behavior, analyzing reasoning capabilities, and critically assessing information.

For example, in a science class, a teacher can ask students to generate questions about a specific topic, such as the human digestive system, that they believe the AI might not have been explicitly trained on. Students can then use ChatGPT to obtain responses and evaluate the accuracy and depth of the AI-generated answers. This activity encourages students to think critically about the information provided by the AI and helps them understand the AI's limitations in processing complex or novel information.

Challenges: While zero-shot prompting can be a powerful educational tool, it is essential to be aware of its challenges and limitations. Some of these challenges include:

1. Potential inaccuracies or "hallucinations" in AI-generated responses, where the AI might provide plausible but incorrect information.
2. Variability in the effectiveness of zero-shot learning across different subject areas and task types.
3. Ethical considerations surrounding the use of AI responses in decision-making and the need for critical evaluation of AI-generated information.

Few-Shot Prompting

Few-shot prompting is a technique that enables AI models to learn from a small number of examples, typically between 1 and 10, to perform a specific task. Unlike zero-shot prompting, which relies solely on the model's general knowledge, few-shot prompting provides the AI with a few examples to guide its responses.

In few-shot prompting, the AI model is given a prompt with a small set of examples demonstrating the desired input-output behavior. The model then learns from these examples to respond to new, unseen inputs. This approach allows for more targeted and context-specific responses compared to zero-shot prompting.

K-12 Application: In a K-12 setting, few-shot prompting can be used to teach students how to provide clear instructions and examples when working with AI. By experiencing the impact of well-crafted examples on AI-generated responses, students can develop a deeper understanding of how AI learns and adapts to specific tasks.

For example, in a language arts class, a teacher can ask students to use few-shot prompting to generate short stories. Students can provide the AI with a few example stories that demonstrate a specific writing style, genre, or theme. The AI will then generate new stories based on these examples, allowing students to analyze how the provided examples influence the AI's output.

WORKSHEETS

Decomposed Prompting Worksheet: Analyzing Character Development

Novel: "To Kill a Mockingbird" by Harper Lee

Main Prompt:

Let's analyze the character development in the novel "To Kill a Mockingbird" by Harper Lee.

Decomposed Prompts:

1. Identify the main characters in the novel.
2. Describe the characters' personalities and beliefs at the beginning of the story.
3. Identify key events that impact each character's development throughout the novel.
4. Analyze how each character's thoughts, actions, and relationships change due to these events.
5. Discuss the lessons each character learns and how they grow by the end of the story.
6. Reflect on how the character development contributes to the overall themes and messages of the novel.

Instructions:

1. Answer each sub-question thoroughly, using evidence from the text to support your responses.
2. Share your responses in small groups or with the class, fostering discussion and collaboration.
3. Synthesize your responses to the sub-questions to create a comprehensive analysis of character development in the novel.
4. Reflect on the process and discuss how breaking down the main question into smaller steps helped you better understand and analyze the text.

Student Responses:

1. Main characters:
2. Characters' initial personalities and beliefs:
3. Key events impacting character development:
4. Changes in characters' thoughts, actions, and relationships:
5. Lessons learned and character growth:
6. Contribution of character development to overall themes and messages:

Reflection:

How did breaking down the main question into smaller steps help you better understand and analyze the text?

Step-by-Step Prompting Worksheet: Multi-Step Math Problems

Problem:

Sarah has 42 stickers. She gives 17 stickers to her friend Tom and then buys 23 more stickers at the store. How many stickers does Sarah have now?

Step-by-Step Prompts:

1. Identify the number of stickers Sarah has at the beginning.
2. Determine how many stickers Sarah gives to Tom.

3. Calculate how many stickers Sarah has left after giving some to Tom.

4. Identify the number of stickers Sarah buys at the store.

5. Add the number of stickers Sarah has left and the number of stickers she buys to find the total number of stickers she has now.

Instructions:

1. Read the problem carefully and identify the important information.

2. Follow the step-by-step prompts to solve the problem.

3. Show your work in the space provided.

4. Write your final answer in a complete sentence.

Student Worksheet:

1. Number of stickers Sarah has at the beginning: _____

2. Number of stickers Sarah gives to Tom: _____

3. Number of stickers Sarah has left after giving some to Tom: ____
   ○ Work: _____

4. Number of stickers Sarah buys at the store: _____

5. Total number of stickers Sarah has now: ____
   ○ Work: _____

Final Answer:

Reflection:

How did following the step-by-step prompts help you solve the problem?

Chain of Thought Prompting Worksheet: Critical Reasoning

Scenario:

You are a detective investigating a mysterious case. A valuable painting has been stolen from the city museum, and you have three suspects: the security guard, the museum curator, and a visiting artist. You have gathered the following clues:

• The security guard was on duty the night the painting was stolen but claims to have seen nothing unusual.

- The museum curator had access to the painting's location and had recently argued with the museum's board about the painting's value.
- The visiting artist was seen near the museum on the night of the theft and had previously expressed admiration for the stolen painting.

Instructions:

1. Read the scenario carefully and identify the important information.
2. Use the Chain of Thought Prompting steps below to guide your reasoning and solve the case.
3. Write your reasoning for each step in the space provided.
4. Present your conclusion and explain your reasoning behind it.

Chain of Thought Prompting Steps:

1. Identify the main suspects and their potential motives.
   - Reasoning: _____
2. Analyze the clues related to each suspect.
   - Security guard: _____
   - Museum curator: _____
   - Visiting artist: _____
3. Consider the reliability of each clue and its relevance to the case.
   - Reasoning: _____
4. Determine which suspect is the most likely culprit based on the available evidence.
   - Reasoning: _____

Conclusion:

Reflection:

How did the Chain of Thought Prompting steps help you organize your reasoning and solve the case?

Multi-Step Prompting Worksheet: Analyzing Historical Events

Topic: The American Revolution

Instructions:

1. Read the provided information about the American Revolution.
2. Follow the multi-step prompts below to analyze the event and its significance.
3. Write your responses in the space provided for each prompt.

Multi-Step Prompts:

1. Identify the main causes of the American Revolution.
   ○ Response: _____
2. Describe the key events that led to the outbreak of the war.
   ○ Response: _____
3. Analyze the roles of significant individuals involved in the American Revolution.
   ○ Response: _____
4. Discuss the impact of the American Revolution on American society and politics.
   ○ Response: _____
5. Evaluate the global significance of the American Revolution and its influence on other countries.
   ○ Response: _____

Reflection:

How did following the multi-step prompts help you gain a deeper understanding of the American Revolution?

Hierarchical Prompting Worksheet: Analyzing an Artistic Movement

Topic: Impressionism

Instructions:

1. Read the provided information about Impressionism.
2. Follow the hierarchical prompts below to analyze the artistic movement and its significance.

3. Write your responses in the space provided for each prompt.

Hierarchical Prompts:

1. Define Impressionism and identify its key characteristics.
   ○ Response: _____
2. Describe the historical and cultural context in which Impressionism emerged.
   ○ Response: _____
3. Analyze the techniques and styles used by prominent Impressionist artists.
   ○ Response: _____
4. Discuss the impact of Impressionism on the art world and its influence on subsequent artistic movements.
   ○ Response: _____
5. Evaluate the lasting significance of Impressionism and its relevance to contemporary art.
   ○ Response: _____

Reflection:

How did the hierarchical organization of the prompts help you develop a comprehensive understanding of Impressionism?

Contrastive Prompting Worksheet: Exploring Literary Devices

Objective: This worksheet is designed to help students understand and differentiate between literary devices using Contrastive Prompting.

Instructions:

1. Read the provided examples of literary devices.
2. Analyze the contrasting elements in each pair of examples.
3. Identify the specific literary device used in each example.
4. Explain how the contrasting elements highlight the differences between the literary devices.

Contrasting Examples:

a. "The sun was a golden orb in the sky, casting its warm rays upon the earth."
b. "The sun, like a giant orange, hung in the sky, its juice dripping down on the world below."

a. "The leaves danced in the wind, their colors a vibrant display of nature's beauty."
b. "The leaves were dancers, twirling and leaping in the wind, their costumes a kaleidoscope of colors."

a. "The old man walked slowly, his cane tapping on the sidewalk with each step."
b. "The old man crept along the sidewalk, his cane a third leg supporting his frail body."

a. "The stars twinkled in the night sky, like diamonds scattered across a black velvet cloth."
b. "The stars winked at me from the night sky as if sharing a secret joke."

Student Worksheet:

- Literary Device: _____
- Literary Device: _____
- Explanation of Contrast: _____

- Literary Device: _____
- Literary Device: _____
- Explanation of Contrast: _____

- Literary Device: _____
- Literary Device: _____
- Explanation of Contrast: _____

- Literary Device: _____
- Literary Device: _____
- Explanation of Contrast: _____

Reflection: How did the contrasting examples help you better understand and differentiate between the literary devices?

Multimodal Prompting Worksheet: Exploring the Water Cycle

Objective: This worksheet is designed to help students understand the water cycle using multimodal prompting and AI-assisted learning.

Instructions:

1. Study the provided image of the water cycle.
2. Read each text-based prompt carefully.
3. Generate your responses using an AI tool like ChatGPT 4 or Claude, incorporating information from the image and your own knowledge.
4. Share your responses with your classmates and discuss the different stages of the water cycle.

[Insert an image of the water cycle here]

Text-based Prompts:

1. Describe the process of evaporation in the water cycle. How does the image illustrate this process?
2. Explain the role of condensation in the water cycle. What visual elements in the image represent condensation?
3. Discuss the importance of precipitation in the water cycle. How does the image depict precipitation?
4. Analyze the process of collection in the water cycle. What components of the image show water collection?

Student Responses:

1. Evaporation:
2. Condensation:
3. Precipitation:
4. Collection:

Reflection: How did the combination of visual aids and text-based prompts help you better understand the water cycle? Did using an AI tool to generate responses enhance your learning experience?

Few-Shot Prompting Worksheet: Generating Short Stories

Objective: This worksheet uses few-shot prompting to help students understand the impact of examples on AI-generated content.

Instructions:

1. Choose a specific writing style, genre, or theme for your short story.
2. Write 3-5 short example stories (50-100 words each) that demonstrate the chosen style, genre, or theme.
3. Use an AI tool like ChatGPT and provide the example stories as a prompt.
4. Generate a new short story using the AI tool.
5. Analyze how the example stories influenced the AI-generated story.

Example Stories:

1. 2. 3. 4. 5.

AI-Generated Story:

Analysis:

1. How did the example stories influence the AI-generated story?
2. What elements of the chosen style, genre, or theme are present in the AI-generated story?
3. How can you refine the example stories to improve the AI-generated output?

Reflection: How does providing examples impact the quality and relevance of AI-generated content compared to zero-shot prompting? What are the benefits and limitations of few-shot prompting in creative writing?

198 | FEARLESSLY USE AI IN THE CLASSROOM

ZERO-SHOT PROMPTING FOR TEACHING CRITICAL THINKING

Objective: This guide is designed to help educators effectively teach critical thinking skills using zero-shot prompting with platforms like ChatGPT. Below is a structured activity you can use as a student handout.

Introduction to Zero-Shot Learning: Zero-shot learning enables AI like ChatGPT to respond to queries based on general knowledge without specific prior examples from the task domain. This capability allows for rapid responses to a wide range of questions, making it a valuable tool in educational settings.

Activity Outline:

1. **Question Formulation:**
   - **Task:** Write down a complex question you want to explore. The question should be something you think ChatGPT might not have been explicitly trained on.
   - **Purpose:** This encourages you to think critically about what information is less likely to be directly available to AI.
2. **Prediction and Analysis:**
   - **Task:** Predict what kind of answer you expect from ChatGPT and why. Then, use ChatGPT to get the actual answer.
   - **Purpose:** This step teaches you to anticipate AI behavior and analyze its reasoning capabilities.
3. **Evaluation of AI Response:**
   - **Task:** Evaluate the accuracy and depth of ChatGPT's response. Consider any potential inaccuracies or "hallucinations" where the AI might provide plausible but incorrect information.
   - **Purpose:** Develop your skills in critically assessing information, highlighting the importance of verification.
4. **Discussion on Limitations:**
   - **Task:** Discuss any observed limitations in ChatGPT's response, especially concerning handling specific data or contexts.
   - **Purpose:** Enhances your understanding of the AI's limitations in processing complex or novel information.
5. **Comparison Across Domains:**
   - **Task:** Repeat the process with different questions from various domains, such as chemistry, genetic counseling, and design education.

- ○ **Purpose**: This will help you see how the effectiveness of zero-shot learning can vary by subject area and task type.

6. **Reflection and Ethical Considerations:**
   - ○ **Task**: Reflect on the ethical implications of using AI responses in decision-making. Discuss scenarios where reliance on AI might lead to ethical dilemmas.
   - ○ **Purpose**: Encourages ethical reasoning and considers how AI should be integrated into learning and decision-making processes.

Conclusion

- **Task**: Summarize your findings and insights from the activities. Discuss how zero-shot learning can be a powerful tool for education and why it's crucial to remain critical of the information provided by AI.

How to Use This Guide:

- Distribute this handout at the beginning of a related lesson or project.
- Allow students time to work through the activities, ideally in groups, to foster discussion.
- Conclude with a class discussion to share findings and reflections.

THANK YOU FOR JOINING THIS JOURNEY

Thank you for choosing this book and for joining us on this exciting exploration of AI in education. Your time, energy, and dedication to improving your teaching practice are truly appreciated. **You are leading the way**, not only for your students but for the future of education as a whole.

If this book has inspired you, provided valuable insights, or helped you rethink the role of AI in your classroom, we would be deeply grateful if you would take a moment to share your thoughts by leaving a **positive review**. Your feedback not only helps others discover this book but also encourages us to continue creating resources that empower educators like you to succeed.

Click here or scan the QR Code below to leave your review on Amazon.

Together, we can redefine what's possible in education and create classrooms where human ingenuity and AI work hand in hand to prepare students for a brighter, more innovative future.

Thank you for being part of this transformative journey. The future begins with you, and it's brighter than ever.

> *"In the end, the power of AI lies not in what it can do for us, but in what we can do with it. The classroom of tomorrow belongs to those who shape it today, let's build it together."*

CONCLUSION

A FUTURE SHAPED TOGETHER: YOUR ROLE IN THE AI-POWERED CLASSROOM

As we reach the final pages of this book, we hope you are leaving with more than just strategies, frameworks, and practical tools for integrating AI into your classroom. We hope you're leaving with a **renewed sense of possibility**, a belief that you, as an educator, are at the forefront of shaping a future where **artificial intelligence enhances, empowers, and inspires learning** in ways that are deeply human.

This journey has explored the transformative power of AI across disciplines, from the intricacies of collaborative problem-solving in STEM to the boundless creative possibilities in arts and humanities and the profound impact of **prompt engineering** on fostering critical thinking, reflection, and personalization. Throughout these pages, you've seen how AI can act as a teammate, a thought partner, and a spark for original ideas, **not a replacement for educators but a powerful extension of your teaching skills**.

At the heart of this transformation lies the most important element of all: **you**. AI is simply a tool. Its success in education depends on the creativity, care, and ethical leadership of educators who guide its use. By taking the bold step to embrace AI, you are not only preparing your students for an AI-driven future but also **modeling the curiosity, adaptability, and critical thinking** they will need to thrive in it.

We've seen how AI can empower less confident students to express their ideas, challenge more advanced learners to think deeper and create dynamic classrooms where **collaboration, creativity, and critical inquiry** thrive. Yet, beyond all the technological advancements, your work as an educator remains the most irreplaceable force of all, connecting knowledge to meaning, tools to purpose, and students to their own potential.

As you step forward to bring these ideas into your classroom, know that your efforts will resonate far beyond the walls of your school. By fostering ethical AI literacy, encouraging innovation, and championing human creativity alongside technological progress, you are shaping a generation of learners who will approach AI not just as consumers but as thoughtful **creators, collaborators, and critical thinkers**.

REFERENCES

https://ciudadaniadigital.mineduc.cl/wp-content/uploads/2023/05/Guia-para-Docentes-Como-usar-ChatGPT-Mineduc.pdf

Fidan, M., & Gencel, N. (2022). Supporting the Instructional Videos With Chatbot and Peer Feedback Mechanisms in Online Learning: The Effects on Learning Performance and Intrinsic Motivation. *Journal of Educational Computing Research, 60*(8), 1716-1741.

Burgess, M. (2009). Using WebCT as a Supplemental Tool to Enhance Critical Thinking and Engagement among Developmental Reading Students. *Journal of College Reading and Learning, 39*(2), 33-49.

Megahed, F., Chen, Y.-J., Ferris, J. A., Knoth, S., & Jones-Farmer, L. A. (2023). How Generative AI models such as ChatGPT can be (Mis)Used in SPC Practice, Education, and Research? An Exploratory Study. *ArXiv.*

Magnisalis, I., Demetriadis, S., & Karakostas, A. (2011). Adaptive and Intelligent Systems for Collaborative Learning Support: A Review of the Field. *IEEE Transactions on Learning Technologies, 4*(1), 5-20.

Karaali, G. (2023). Artificial Intelligence, Basic Skills, and Quantitative Literacy. *Numeracy.*

Adebusuyi, O., Bamidele, E., & Adebusuyi, A. (2020). Effects of in-service chemistry teachers' technological pedagogical content knowledge on students' scientific attitude and literacy in southwestern Nigerian secondary schools. European Journal of Interactive Multimedia and Education, 1(2), e02009. https://doi.org/10.30935/ejimed/9306

Allen, L. K., & Kendeou, P. (2024). ED-AI Lit: An Interdisciplinary Framework for AI Literacy in Education. Policy Insights from the Behavioral and Brain Sciences. SAGE Publications. DOI: 10.1177/23727322231220339

Alsobeh, A., & Woodward, B. (2023). AI as a Partner in Learning: A Novel Student-in-the-Loop Framework for Enhanced Student Engagement and Outcomes in Higher Education. In Proceedings of the 24th Annual Conference on Information Technology Education (SIGITE '23). Association for Computing Machinery, New York, NY, USA, 171–172. https://doi.org/10.1145/3585059.3611405

Astari, M., Masykuri, M., Susilowati, E., & Yamtinah, S. (2023). Effectiveness of online learning with the web-based TPACK scaffolding for enhancement TPACK ability of pre-service chemistry teachers. Journal of Curriculum and Teaching, 12(1), 183. https://doi.org/10.5430/jct.v12n1p183

Binjha, P., Das, B., Dansana, A., & Das, B. (2023). Pedagogical innovation of TPACK based K-4 learning transaction model in science and social sciences. Asian Journal of Education and Social Studies, 39(1838), 42-52. https://doi.org/10.9734/ajess/2023/v39i1838

Celik, I. (2023). Towards Intelligent-TPACK: An empirical study on teachers' professional knowledge to ethically integrate artificial intelligence (AI)-based tools into education. Computers in Human Behavior, 138, 107468. https://doi.org/10.1016/j.chb.2022.107468

Code.org, CoSN, Digital Promise, European EdTech Alliance, & PACE. (n.d.). AI Guidance for Schools Toolkit.

Ginting, D. & Linarsih, A. (2022). Teacher professional development in the perspective of technology pedagogical content knowledge theoretical framework. Jurnal Visi Ilmu Pendidikan, 14(1), 1. https://doi.org/10.26418/jvip.v14i1.49334

Gunanto, Y. & Supriyadi, L. (2021). A case study: technological pedagogical and content knowledge (TPACK) of pre-service physics teacher to enhance the 4C's skills during online learning. Jurnal Penelitian Pendidikan IPA, 7(4), 660-668. https://doi.org/10.29303/jppipa.v7i4.789

Juanda, A., Shidiq, A., & Nasrudin, D. (2021). Teacher learning management: investigating biology teachers' TPACK to conduct learning during the COVID-19 outbreak. Jurnal Pendidikan Ipa Indonesia, 10(1), 48-59. https://doi.org/10.15294/jpii.v10i1.26499

Nevrita, N., Oprasmani, E., & Sarkity, D. (2022). Validity of technological pedagogical content knowledge (TPACK) instruments in learning media for science and biology teachers. EAI Endorsed Transactions on Learning Technologies, 16(11), Article 2326124. https://doi.org/10.4108/eai.16-11-2022.2326124

Paidi, P., Yuniawan, A., & Luthfiyah, F. (2020). Integrating technology in teaching: How teachers use TPACK in education. Journal of Educational and Social Research, 10(1), 31-40. https://doi.org/10.2478/jesr-2020-0003

Pala, K., Dukmak, S., & Arslan, S. (2020). Effect of integrated TPACK instructional approach on teachers' perceptions of technology integration.

Akgün, S., & Greenhow, C. (2021). Artificial intelligence in education: Addressing ethical challenges in K-12 settings. *AI and Ethics*, 2(3), 431-440. https://doi.org/10.1007/s43681-021-00096-7

Bogojević, J., & Pance, B. (2022). Musical creativity in the teaching practice in Montenegrin and Slovenian primary schools. *British Journal of Music Education*, 39(2), 169-182. https://doi.org/10.1017/s0265051722000018

Chiriacescu, F., Chiriacescu, B., Grecu, A., Miron, C., Panisoara, I., & Lazar, I. (2023). Secondary teachers' competencies and attitude: A mediated multigroup model based on usefulness and enjoyment to examine the differences between key dimensions of STEM teaching practice. *Plos One*, 18(1), e0279986. https://doi.org/10.1371/journal.pone.0279986

Dai, Y. (2023). Effect of an analogy-based approach of artificial intelligence pedagogy in upper primary schools. *Journal of Educational Computing Research*, 61(8), 159-186. https://doi.org/10.1177/07356331231201342

Ihnatova, O., Poseletska, K., Matiiuk, D., Hapchuk, Y., & Borovska, O. (2021). Application of digital technologies in teaching a foreign language in a blended learning environment. *Linguistics and Culture Review*, 5(S4), 114-127. https://doi.org/10.21744/lingcure.v5ns4.1571

Liu, Y. (2022). Application of artificial intelligence algorithm and VR technology in vocal music teaching. *Mobile Information Systems*, 2022, 1-13. https://doi.org/10.1155/2022/2320198

Mahaffey, A. (2020). Chemistry in a cup of coffee: Adapting an online lab module for teaching specific heat capacity of beverages to health sciences students during the COVID pandemic. *Biochemistry and Molecular Biology Education*, 48(5), 528-531. https://doi.org/10.1002/bmb.21439

Razali, H., Jamaluddin, R., & Kamarudin, N. (2022). Implementing integrated STEM teaching in design and technology: Teachers' knowledge and teaching practices. *International Journal of Academic Research in Business and Social Sciences*, 12(9). https://doi.org/10.6007/ijarbss/v12-i9/14785

Roehrig, G., Ellis, J., & Dare, E. (2021). Beyond the basics: A detailed conceptual framework of integrated STEM. *Disciplinary and Interdisciplinary Science Education Research*, 3(1). https://doi.org/10.1186/s43031-021-00041-y

Shaheen, M. (2021). Self-determination theory for motivation in distance music education. *Journal of Music Teacher Education*, 31(2), 80-91. https://doi.org/10.1177/10570837211062216

Tawbush, R., Stanley, S., Campbell, T., & Webb, M. (2020). International comparison of K-12 STEM teaching practices. *Journal of Research in Innovative Teaching & Learning*, 13(1), 115-128. https://doi.org/10.1108/jrit-01-2020-0004

Tsoukala, C. (2021). STEM integrated education and multimodal educational material. *Advances in Mobile Learning Educational Research*, 1(2), 96-113. https://doi.org/10.25082/amler.2021.02.005

Turner, M. (2020). Drawing on students' diverse language resources to facilitate learning in a Japanese–English bilingual program in Australia. *Language Teaching Research*, 25(1), 61-80. https://doi.org/10.1177/1362168820938824

Xiao, H. (2022). Innovation of digital multimedia VR technology in music education curriculum in colleges and universities. *Scientific Programming*, 2022, 1-9. https://doi.org/10.1155/2022/6566144

Xu, Y., & Nazir, S. (2022). Ranking the art design and applications of artificial intelligence and machine learning. *Journal of Software Evolution and Process, 36*(2). https://doi.org/10.1002/smr.2486

Zhou, D., Liu, Y., Huang, J., Xiang, Y., Gu, R., & Liu, B. (2023). An intelligent tutoring system enhancing transdisciplinary problem-finding in design-led integrated STEM education. https://doi.org/10.2991/978-94-6463-040-4_142

Bellas, M. L., Bellas, M. D., & Díaz, V. D. (2022). Interdisciplinary approach to improve educational outcomes in higher education: A systematic literature review. Journal of Interdisciplinary Education, 11(2), 102-120. https://doi.org/10.1080/21682634.2022.2045698

Calatayud, V., Espinosa, M., & Vila, R. (2021). Artificial intelligence for student assessment: A systematic review. Applied Sciences, 11(12), 5467. https://doi.org/10.3390/app11125467

Elzain, M., Moran, L., McCarth, G., Hyde, S., & McFarland, J. (2022). Evaluation of postgraduate educational environment of doctors training in psychiatry: A mixed method study. Journal of Medical Education, 9(3), 78-92. https://doi.org/10.1101/2022.02.24.481497

Emre, İ., & Çelik, M. (2021). Determining the technological and pedagogical content knowledge level of 4th-grade teachers on the unit; electric in our life. Turkish Journal of Educational Studies, 8(1), 1-25. https://doi.org/10.33907/turkjes.813002

Fan, Z., Zhang, J., Xie, J., & Zhang, L. (2021). Machine learning and artificial intelligence in education: A review. IEEE Access, 9, 44927-44940. https://doi.org/10.1109/access.2021.3067352

Google for Education. (2022). Google classroom. https://edu.google.com/products/classroom/

Huang, Y., Tang, C., & Liu, X. (2021). Artificial intelligence in education: Technologies, applications and trends. IEEE Transactions on Learning Technologies, 14(4), 783-789. https://doi.org/10.1109/tlt.2021.3097501

IBM. (2021). IBM Watson education. https://www.ibm.com/watson/education

Karaca, E., Başbay, M., & Yilmaz, K. (2021). The importance of the use of artificial intelligence applications in education. In A. T. Tsalakanidou (Ed.), Proceedings of the 2nd International Conference on Education and E-Learning (ICEEL 2021)

(pp. 32-37). https://doi.org/10.1145/3472703.3472768

Kausik, S., Jordon, M., & Goldberg, L. (2022). Artificial intelligence in education: A bibliometric analysis. Educational Technology Research and Development, 70(3), 1295-1318. https://doi.org/10.1007/s11423-022-10045-5

Kim, Y., Monroe, E., Nielsen, H., Cox, J., Southard, K., Elfring, L., ... & Talanquer,

V. (2022). Exploring undergraduate students' abilities to collect and interpret formative assessment data. Journal of Chemical Education, 99(3), 1410-1419. https://doi.org/10.1021/acs.jchemed.1c01145

Ahimaz, P. (2023). Genetic counselors' utilization of chatgpt in professional practice: a cross-sectional study. *American Journal of Medical Genetics Part A*, 194(4). https://doi.org/10.1002/ajmg.a.63493

Ali, S. (2023). General purpose large language models match human performance on gastroenterology board exam self-assessments. https://doi.org/10.1101/2023.09.21.23295918

Araújo, J. (2024). Can chatgpt enhance chemistry laboratory teaching? using prompt engineering to enable ai in generating laboratory activities. *Journal of Chemical Education*. https://doi.org/10.1021/acs.jchemed.3c00745

Biswas, S. (2024). Utility of artificial intelligence-based large language models in ophthalmic care. *Ophthalmic and Physiological Optics*, 44(3), 641-671. https://doi.org/10.1111/opo.13284

Cheng, D., Huang, S., Bi, J., Zhan, Y., Liu, J., Wang, Y., ... & Zhang, Q. (2023). Uprise: universal prompt retrieval for improving zero-shot evaluation. https://doi.org/10 48550/arxiv.2303.08518

Gu, J., Tresp, V., & Qin, Y. (2022). Are vision transformers robust to patch perturbations?, 404-421. https://doi.org/10.1007/978-3-031-19775-8_24

Kashihara, K., Pal, K., Baral, C., & Trevino, R. (2023). Prompt-based learning for thread structure prediction in cybersecurity forums. https://doi.org/10.48550/arxiv.2303.05400

Leon, A. (2023). Chatgpt needs a chemistry tutor too. *Journal of Chemical Education*, 100(10), 3859-3865. https://doi.org/10.1021/acs.jchemed.3c00288

Liu, P., Yuan, W., Fu, J., Jiang, Z., Hayashi, H., & Neubig, G. (2021). Pre-train, prompt, and predict: a systematic survey of prompting methods in natural language processing. https://doi.org/10.48550/arxiv.2107.13586

Lund, B. (2023). The prompt engineering librarian. *Library Hi Tech News*, 40(8), 6-8. https://doi.org/10.1108/lhtn-10-2023-0189

Meron, Y. (2023). Artificial intelligence in design education: evaluating chatgpt as a virtual colleague for post-graduate course development. *Design Science*, 9. https://doi.org/10.1017/dsj.2023.28

Polverini, G. (2024). How understanding large language models can inform the use of chatgpt in physics education. *European Journal of Physics*, 45(2), 025701. https://doi.org/10.1088/1361-6404/ad1420

Russe, M. (2024). Improving the use of llms in radiology through prompt engineering: from precision prompts to zero-shot learning. *Röfo - Fortschritte Auf Dem Gebiet Der Röntgenstrahlen Und Der Bildgebenden Verfahren*. https://doi.org/10.1055/a-2264-5631

Sorensen, T., Robinson, J., Rytting, C., Shaw, A., Rogers, K., Delorey, A., … & Wingate, D. (2022). An information-theoretic approach to prompt engineering without ground truth labels. https://doi.org/10.18653/v1/2022.acl-long.60

White, A., Hocky, G., Gandhi, H., Ansari, M., Cox, S., Wellawatte, G., … & Ccoa, W. (2022). Assessment of chemistry knowledge in large language models that generate code. https://doi.org/10.26434/chemrxiv-2022-3md3n-v2

Wilkes, T., Stark, L., Trempler, K., & Stark, R. (2022). Contrastive video examples in teacher education: a matter of sequence and prompts. *Frontiers in Education*, 7. https://doi.org/10.3389/feduc.2022.869664

Xu, Z., Wang, C., Qiu, M., Luo, F., Xu, R., Huang, S., … & Huang, J. (2022). Making pre-trained language models end-to-end few-shot learners with contrastive prompt tuning. https://doi.org/10.48550/arxiv.2204.00166

Yong, G., Jeon, K., Gil, D., & Lee, G. (2022). Prompt engineering for zero-shot and few-shot defect detection and classification using a visual-language pretrained model. *Computer-Aided Civil and Infrastructure Engineering*, 38(11), 1536-1554. https://doi.org/10.1111/mice.12954

Zhang, H., Li, D., Li, Y., Shang, C., Shi, C., & Jiang, Y. (2023). Assisting language learners: automated trans-lingual definition generation via contrastive prompt learning. https://doi.org/10.18653/v1/2023.bea-1.23

King, J., Kim, S., Smith, T., & Jones, L. (2022). Leveraging artificial intelligence for personalized learning: Opportunities and challenges. Journal of Educational Technology & Society, 25(1), 237-247. https://www.jstor.org/stable/jeductechsoci.25.1.237

Laat, M. F., & Joksimovic, S. (2020). Challenges of workplace learning analytics: Ethical issues and stakeholders' perspectives. Proceedings of the Tenth International Conference on Learning Analytics & Knowledge (LAK '20) (pp. 435- 444). https://doi.org/10.1145/3375462.3375510

Lester, J. C. (2024). Artificial intelligence for education: A comprehensive overview. Cambridge University Press.

Lin, L. (2020). Investigating the role of artificial intelligence in education: A literature review. International Journal of Artificial Intelligence in Education, 30(4), 470-497. https://doi.org/10.1007/s40593-019-00202-7

Memmert, D., & Bittner, A. C. (2022). Testing the efficacy of a basketball-specific cognitive-motor dual-task training program. International Journal of Environmental Research and Public Health, 19(4), 2288. https://doi.org/10.3390/ijerph19042288

Merchie, E., Gijbels, D., Donche, V., & Van den Bossche, P. (2018). The impact of formative assessment on self-regulated learning and self-efficacy beliefs in writing: A longitudinal study. Metacognition and Learning, 13(3), 303-321. https://doi.org/10.1007/s11409-018-9172-x

Mlodzinski, E., Wardi, G., Viglione, C., Nemati, S., Alexander, L., & Malhotra, A. (2023). Assessing barriers to implementation of machine learning and artificial intelligence–based tools in critical care: Web-based survey study. JMIR Perioperative Medicine, 6, e41056. https://doi.org/10.2196/41056

Murtaza, M., Ahmed, Y., Shamsi, J., Sherwani, F., & Usman, M. (2022). AI-based personalized e-learning systems: Issues, challenges, and solutions. IEEE Access, 10, 81323-81342. https://doi.org/10.1109/access.2022.3193938

Nağaç, M., & Kalayci, S. (2021). The effect of STEM activities on students' academic achievement and problem-solving skills: Matter and heat unit. E-Kafkas Eğitim Araştırmaları Dergisi, 8(3), 480-498. https://doi.org/10.30900/kafkasegt.964063

Nazaretsky, J., Milford, T., & Monahan, J. (2021). AI in education: A look back, a look forward. EDUCAUSE Review, 56(4), 34-44. https://er.educause.edu/articles/2021/8/ai-in-education-a-look-back-a-look-forward

Nixon, M. (2024). Artificial intelligence in education: Recent advances and future directions. Educational Technology & Society, 27(1), 312-325. https://www.jstor.org/stable/26993914

Rushton, E., Nixon, H., & Schwarz, L. (2022). Ethical considerations in artificial intelligence-driven education: A scoping review. Journal of Information Technology Education: Research, 21, 375-407. https://doi.org/10.28945/5078

Schilling, J., Moeller, F., Peterson, R., Beltz, B., Joshi, D., Gartner, D., ... & Jain, P. (2023). Testing the acceptability and usability of an AI-enabled COVID-19 diagnostic tool among diverse adult populations in the United States. Quality Management in Health Care, 32(Supplement 1), S35-S44. https://doi.org/10.1097/qmh.0000000000000396

Touretzky, D. S., Newsome, W. T., Hasselmo, M. E., & Loeb, G. E. (2019). Using neuromorphic hardware to explore brain-inspired learning. Frontiers in Neuroscience, 13, 166. https://doi.org/10.3389/fnins.2019.00166

Trovato, G., & Russo, M. (2021). Artificial intelligence (AI) and lung ultrasound in infectious pulmonary disease. Frontiers in Medicine, 8. https://doi.org/10.3389/fmed.2021.706794

Zahra, F., Keong, L., & Ismail, N. (2022). Implementation of artificial intelligence in education: A review of case studies. Journal of Educational Technology & Society, 25(3), 168-182. https://www.jstor.org/stable/26993914

Akgün, S., & Greenhow, C. (2021). Artificial intelligence in education: addressing ethical challenges in K-12 settings. AI and Ethics, 2(3), 431-440. https://doi.org/10.1007/s43681-021-00096-7

Anantrasirichai, N., & Bull, D. (2022). Artificial intelligence in the creative industries: a review. Artificial Intelligence Review, 55, 589–656. https://doi.org/10.1007/s10462-021-10039-7

Gadanidis, G. (2024). Mathematics & artificial intelligence: intersections and educational implications. Journal of Digital Life and Learning, 4(1), 1-24. https://doi.org/10.51357/jdll.v4i1.249

Gunser, V., Gottschling, S., Brucker, B., Richter, S., Çakir, D., & Gerjets, P. (2022). The pure poet: how good is the subjective credibility and stylistic quality of literary short texts written with an artificial intelligence tool as compared to texts written by human authors? https://doi.org/10.18653/v1/2022.in2writing-1.8

Holmes, W., Porayska-Pomsta, K., Holstein, K., Sutherland, E., Baker, T., Shum, S., ... & Koedinger, K. (2021). Ethics of AI in education: towards a community-wide framework. International Journal of Artificial Intelligence in Education, 32(3), 504-526. https://doi.org/10.1007/s40593-021-00239-1

Krakowski, A., Greenwald, E., Hurt, T., Nonnecke, B., & Cannady, M. (2022). Authentic integration of ethics and AI through sociotechnical, problem-based learning. Proceedings of the AAAI Conference on Artificial Intelligence, 36(11), 12774-12782. https://doi.org/10.1609/aaai.v36i11.21556

Rezwana, J., & Maher, M. L. (2023). Designing creative AI partners with COFI: A framework for modeling inter-action in human-AI co-creative systems. ACM Transactions on Computer-Human Interaction, 30(5), Article 67. https://doi.org/10.1145/3519026

Yusa, I., Yu, Y., & Sovhyra, T. (2022). Reflections on the use of artificial intelligence in works of art. JADAM, 2(2), 152-167. https://doi.org/10.58982/jadam.v2i2.334